sharks, skates, and rays of the carolinas

Frank J. Schwartz

sharks, skates, and rays of the carolinas

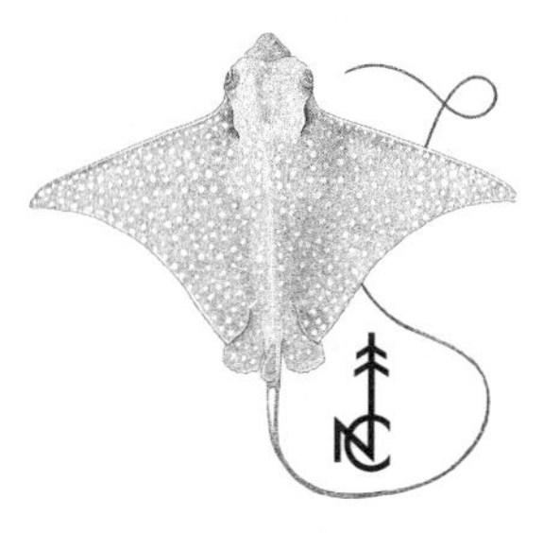

The University of North Carolina Press Chapel Hill and London

Manufactured in the United States of America

Designed by Heidi Perov
Set in Cycles

The paper in this book meets the guidelines for permanence and durability of the Committee on Production Guidelines for Book Longevity of the Council on Library Resources.

Library of Congress Cataloging-in-Publication Data
Schwartz, Frank Joseph, 1929–
Sharks, skates, and rays of the Carolinas / by Frank J. Schwartz.
p. cm.
Includes bibliographical references and index.
ISBN 0-8078-5466-2 (pbk.: alk. paper)
1. Sharks—North Carolina. 2. Sharks—South Carolina. 3. Skates (Fishes)—North Carolina. 4. Skates (Fishes)—South Carolina. 5. Rays (Fishes)—North Carolina. 6. Rays (Fishes)—South Carolina. I. Title.
QL638.9.S294 2003
597.3′0975—dc21 2002152604

07 06 05 04 03 5 4 3 2 1

contents

A section of color illustrations appears after p. 116.

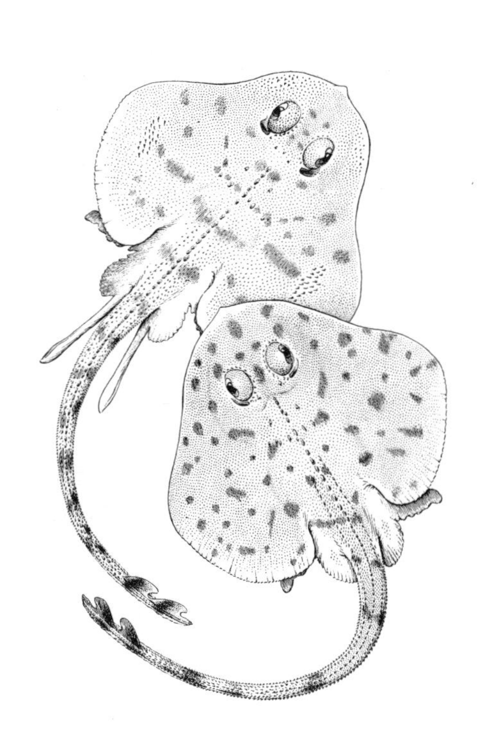

acknowledgments

Many have helped in the field during the past thirty-five years, supplied information, and encouraged the gathering of the data that comprises this volume. The Institute of Marine Sciences, Morehead City, North Carolina, provided constant support in the form of assistance, facilities, and the use of R/V *Machapunga* and R/V *Capricorn*. Institute ship captains were J. Wagoner, T. Kellum, O. Lewis Jr., and J. Purifoy. Mates were J. Spencer, W. Willis, and S. Davis. Research technician was G. Safrit Jr. A host of students followed me to learn the secrets of the sea, especially about sharks, skates, and rays. They also learned that the sea is an interesting place that can be nice one day and a terror another. So too, a day at sea helped weed out oceanographers from those with real interests in zoology. Additional information and specimens were obtained from the entire sport and commercial fleets and fishermen of North Carolina. Knowledge of deepwater elasmobranchs was made possible while on R/V *Eastward* and R/V *Cape Hatteras*, ably assisted by G. Newton, F. Kelly, J. Newton, and C. Olson. A. Powell, J. Sterling, F. Holland, and J. Gillikin, then of the North Carolina Division of Marine Fisheries, Morehead City, North Carolina, provided deep station material collected by R/V *Dan Moore*, captained by F. Smith.

Information regarding the old shark fisheries out of Morehead City was furnished by O. Phillips and about the Ocean Leather Corporation by L. Moresi, Orange, New Jersey. Information about South Carolina elasmobranchs was provided by C. Barans, C. Moore, G. Ulrich, and S. van Sant (now of the Virgin Islands) of the South Carolina Marine Resources Research Institute, Charleston.

Permission to modify drawings of *Echinorhinus brucus*, *Centroscyllium fabricii*, *Rhincodon typus*, *Lamna nasus*, *Alopias superciliosus*, *Galeus arae*, *Apris-*

turus profundorum, Carcharhinus brevipinna, Squalus cubensis, and *Carcharhinus isodon* from volume 1 of *Western North Atlantic fishes* was granted by the late Dr. D. Merriman (Sears Foundation Marine Research); *Deania profundorum,* by the late S. Springer; *Carcharhinus altimus,* by Dr. J. Garrick (Victoria University of Wellington, New Zealand); *Prionace glauca,* by the late Dr. W. Templeman (Memorial University, St. Johns, Newfoundland); and *Somniosus microcephalus,* by J. Casey (National Marine fisheries Service, Narragansett, R.I.). Dr. M. Stehmann (Institut für Seefischerei, Hamburg, Germany) commented on the status of *Apristurus laurussoni* and *Apristurus profundorum* and permitted use of *Neoraja carolinensis* drawings published in the *Proceedings of the Biological Society of Washington, D.C.,* edited by Dr. B. Robbins. The late Dr. G. Krefft (Institut für Seefischerei, Hamburg, Germany) provided the loan of the paratype (ISH 1573/66) and a young specimen of *Etmopterus gracilispinis* on which the drawing is based. Dr. G. Gilmore (Vero Beach, Fla.) provided photos and measurements from which *Isurus paucus* was constructed. Dr. J. McEachran (Texas A&M University, College Station) provided a loan of a *Bathyraja richardsoni* specimen; Dr. R. Winterbottom (Royal Ontario Museum, Canada), a *Pteroplatytrygon violacea*; and Dr. V. Springer (U.S. National Museum, Washington, D.C.), a *Centrophorus granulosus* for drawing.

Dr. J. Merriner (National Marine fishery Service, Beaufort, N.C.) granted permission to use S. Springer's drawings of *Apristurus*. D. Flescher provided the slide from which *Heptranchias perlo* was made. Smalltooth sawfish rostrums were provided for measurement by Captain O. Purifoy, J. L. Seamon, and the Hampton Mariners Museum, Beaufort, North Carolina.

R. Barnes (Institute of Marine Sciences, Morehead City, N.C.) redrew the *Apristurus* and added several *Carcharhinus* drawings. Leslie M. V. Murdock (Annapolis, Md.) patiently produced the original accurate line and stippled drawings of the sharks, skates, and rays. L. White deciphered my hieroglyphics and typed the text; G. Safrit Jr. typed the keys and several minor sections. Librarians B. Bright (Institute of Marine Sciences, Morehead City, N.C.) and P. Marraro (National Marine fisheries Service, Beaufort, N.C.) helped with literature retrieval.

I express my appreciation to the University of North Carolina Press at Chapel Hill staff, especially Elaine Maisner, Adrienne Allison, Stephanie

Wenzel, and Paula Wald for encouragement and aid in completing the revision of the 1984 and 1989 publications: *Sharks of the Carolinas* and *Sharks, Sawfish, Skates, and Rays of the Carolinas*. Their suggestions were greatly welcomed. Dr. W. Hamlett (Indiana University School of Medicine) reviewed the section on reproduction. Dr. R. W. Purdy (Department of Paleobiology, Smithsonian Institution, Washington, D.C.) and an anonymous reader reviewed the entire manuscript prior to publication and made important suggestions for improvement. C. Jensen (North Carolina Division of Marine fisheries, Morehead City) helped for many years in the field and kept constantly supplying current information on his shark studies in North Carolina, especially those dealing with neonate stages of sharks. Alan Williamson and especially Raymond Churchill were able retired volunteers who helped out during all kinds of weather and sea conditions.

sharks, skates, and rays of the carolinas

introduction

What was that 4.5-meter-long shadow that swam by during a scuba diving session? Which sharks are considered dangerous and are usually associated with "shark attacks"? Why and when do attacks occur? Are stingrays harmless? What is the difference between a shark, skate, or ray and whales and porpoises? What is the difference between a true blacktip shark and other sharks with black-tipped fins? Do sharks, skates, and rays ever enter estuaries? Freshwater? When and where should they be expected? Are all elasmobranchs always gray? Where can fossil shark teeth, vertebrae, and other parts be found? How do sharks reproduce? Do sharks have parasites? Are they ever deformed? Have sport and commercial fisheries affected the size of shark populations? What is the outlook for sharks and other elasmobranchs?

These are all good questions, but what *do* we really know about sharks, skates, and rays? This volume resolves many of those questions and others concerning the 56 species of sharks and 35 species of sawfish, skates, and stingrays known to occur in the waters off North Carolina and South Carolina (see Table 1).

Fishes, 24,618 known species, comprise more than half of the 48,170 known living vertebrate species of the world (Nelson 1994). Sharks, skates, rays, and sawfish are cartilaginous fishes (class Chondrichthyes), which means that their bodies are supported by cartilaginous skeletons rather than bone. Scientists have organized the 1,164 species of known cartilaginous fishes into 60 families and 185 genera (as of 1999). At least 67 species of sharks and 57 skates and rays remain to be described (Compagno 1999).

The Chondrichthyes are composed of the elasmobranchs (sharks and rays, 96 percent) and the living holocephalans (chimeras and elephant

TABLE 1. *Scientific and common names of 56 sharks and 35 skates and rays known from the western Atlantic Ocean and the inland waters of the Carolinas*

Sharks

SCIENTIFIC NAME	COMMON NAME
Squatina dumeril	Atlantic angel shark
Hexanchus griseus	Bluntnose sixgill shark
Heptranchias perlo	Sharpnose sevengill shark
Ginglymostoma cirratum	Nurse shark
Echinorhinus brucus	Bramble shark
Somniosus microcephalus	Greenland shark
Dalatias licha	Kitefin shark
Deania profundorum	Arrowhead dogfish
Centrophorus granulosus	Gulper shark
Centroscyllium fabricii	Black dogfish
Etmopterus bullisi	Lined lantern shark
Etmopterus gracilispinis	Broadband lantern shark
Etmopterus hillianus	Caribbean lantern shark
Cirrhigaleus asper	Roughskin spurdog
Squalus acanthias	Piked (spiny) dogfish
Squalus cubensis	Cuban dogfish
Squalus mitsukurii	Shortspine spurdog
Sphyrna lewini	Scalloped hammerhead
Sphyrna mokarran	Great hammerhead
Sphyrna zygaena	Smooth hammerhead
Sphyrna tiburo	Bonnethead shark
Carcharias taurus	Sand tiger shark
Odontaspis ferox	Smalltooth sand tiger
Alopias superciliosus	Bigeye thresher
Alopias vulpinus	Thresher shark
Cetorhinus maximus	Basking shark
Carcharodon carcharias	Great white shark
Isurus oxyrinchus	Shortfin mako
Isurus paucus	Longfin mako
Lamna nasus	Porbeagle
Rhincodon typus	Whale shark
Apristurus laurussoni	Iceland cat shark
Apristurus profundorum	Deepwater cat shark
Galeus arae	Roughtail cat shark

Sharks *(continued)*

SCIENTIFIC NAME	COMMON NAME
Scyliorhinus hesperius	White-saddled cat shark
Scyliorhinus meadi	Blotched cat shark
Scyliorhinus retifer	Chain cat shark
Mustelus canis	Dusky smooth hound
Mustelus norrisi	Florida smooth hound
Carcharhinus acronotus	Blacknose shark
Carcharhinus altimus	Bignose shark
Carcharhinus brevipinna	Spinner shark
Carcharhinus falciformis	Silky shark
Carcharhinus galapagensis	Galapagos shark
Carcharhinus isodon	Finetooth shark
Carcharhinus leucas	Bull shark
Carcharhinus limbatus	Blacktip shark
Carcharhinus longimanus	Oceanic whitetip shark
Carcharhinus obscurus	Dusky shark
Carcharhinus perezi	Caribbean reef shark
Carcharhinus plumbeus	Sandbar shark
Carcharhinus signatus	Night shark
Galeocerdo cuvier	Tiger shark
Negaprion brevirostris	Lemon shark
Prionace glauca	Blue shark
Rhizoprionodon terraenovae	Atlantic sharpnose shark

Skates and Rays

SCIENTIFIC NAME	COMMON NAME
Pristis pectinata	Smalltooth sawfish
Rhinobatos lentiginosus	Atlantic guitarfish
Benthobatis marcida	Deep-sea blind ray
Narcine brasiliensis	Lesser electric ray
Torpedo nobiliana	Atlantic torpedo
Dactylobatus armatus	Skilletskate
Dipturus teevani	Caribbean skate
Dipturus laevis	Barndoor skate
Rajella bathyphila	Deepwater skate
Bathyraja richardsoni	Richardson's skate
Fenestraja plutonia	Pluto skate

SCIENTIFIC NAME	COMMON NAME
Breviraja spinosa	Spinose skate
Leucoraja ocellata	Winter skate
Leucoraja erinacea	Little skate
Malacoraja senta	Smooth skate
Neoraja carolinensis	Carolina pygmy skate
Fenestraja atripinna	Blackfin pygmy skate
Amblyraja radiata	Thorny skate
Leucoraja garmani	Rosette skate
Raja eglanteria	Clearnose skate
Urobatis jamaicensis	Yellow stingray
Pteroplatytrygon violacea	Pelagic stingray
Dasyatis sabina	Atlantic stingray
Dasyatis say	Bluntnose stingray
Dasyatis americana	Southern stingray
Dasyatis centroura	Roughtail stingray
Gymnura altavela	Spiny butterfly ray
Gymnura micrura	Smooth butterfly ray
Manta birostris	Manta ray
Mobula hypostoma	Atlantic devil ray
Mobula mobular	Giant devil ray
Rhinoptera bonasus	Cownose ray
Aetobatus narinari	Spotted eagle ray
Myliobatis freminvillii	Bullnose ray
Myliobatis goodei	Southern eagle ray

fishes, 4 percent) (Compagno 1999). Elasmobranchs are composed of two monophyletic assemblages: sharks (Squalomorphii, 491+ species, 44 percent) and skates, rays, and sawfish (a type of batoid ray) (Rajomorphii, 626+ species, 56 percent).

While the elasmobranchs have been studied at length, much remains unresolved (Bigelow and Schroeder 1948, 1953; Choi, Kim, Nakaya 1998; Compagno 1973, 1977, 1984, 1988, 2001; Compagno and Cook 1995; Compagno et al. 1990; de Carvalho 1996; Dingerkus 1986, 1995; Dunn and Morrissey 1995; Garrick 1960b, 1967a, 1967b, 1982; Goto 2001; Hamlett 1999;

Heemstra 1997; Heemstra and Smith 1980; Lovejoy 1996; Maisey 1984, 1986; McEachran 1982; McEachran and Miyake 1990a; Nakaya and Sato 1997; Naylor 1992; Naylor et al. 1997; Nishida 1990; Rosenberger 2001; Schwartz and Maddock 1986, 2002; Shirai 1992a, 1992b, 1996; Stiassny, Parenti, and Johnson 1996; Thies 1987; Zangerl 1973, 1981).

General Description

Elasmobranchs are primitive fishes possessing cartilaginous skeletons (only the small-spotted cat shark, *Scyliorhinus canicula,* has some bone) (Hall 1982) and have 5 to 7 pairs of gill slits not covered by an opercle (a stiff gill-plate cover found in bony fishes). The skin may be fully or partially covered with modified scales called denticles or thorns. All species have internal fertilization, and males possess a pair of elongations of their pelvic fins (claspers) for sperm intromission and insemination of a female.

Sharks and saw sharks are distinguished from sawfishes, skates, and rays by possessing rigid dorsal fins and 5 to 7 gill slits on each side of the head; the pectoral fins are located behind the gill slits and are not attached to the head. An anal fin may be present. The elongate caudal fin may have various shapes. Conversely, sawfishes, skates, and rays, called batoids, have 5 to 6 gill slits on the ventral surface of the head and conspicuous spiracles (openings, for water intake during breathing, located dorsally and laterally behind the eyes; a few species of sharks have visible spiracles). Enlarged pectoral fins are continuous and join the head. A nictitating membrane (false eyelid, present in requiem carcharhinid sharks) is absent. The anal fin is absent, and the tail may or may not be long, stout, equally lobed, or whiplike. Sawfish lack snout rostral barbels, resemble and possess features similar to those of saw sharks that have rostral barbels, and are batoids because their pectoral fins are broadly connected to the head.

Skates and rays are diamond shaped, flattened dorso-ventrally, and except for mantas (*Manta birostris*), butterfly rays (*Gymnura* sp.), southern stingrays (*Dasyatis americana*), and roughtail stingrays (*Dasyatis centroura*), rarely exceed 1-meter wing disk widths (DW). Skates can be further distinguished from stingrays as they possess two dorsal fins on the tail, have a distinct caudal fin, possess thorns on the tail, and lack poisonous serrated

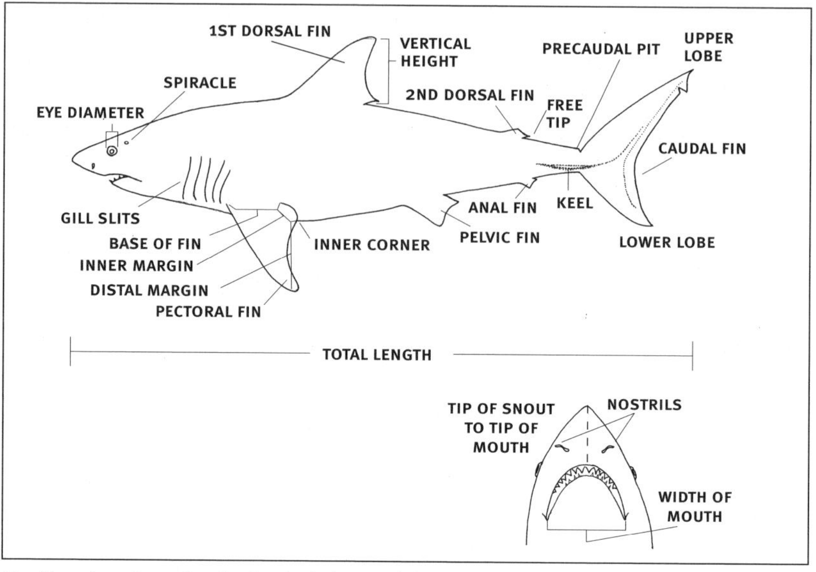

Outline drawing of a shark, slightly modified to include important features.

spines on the tail. Stingrays possess one or no dorsal fins, and their whip-like tails are usually armed with one or more serrated spines that possess venom-producing goblet cells located along the length of ventral grooves of the spine. Tail length may vary from equal to or several times longer than the body length (roughtail stingray). All stingrays have 5 gill slits, except an African species that has 6 (*Hexatrygon bickelli*) (Heemstra and Smith 1980). Elasmobranchs can be distinguished from whales and porpoises as they have gill slits and lack blowholes on the top of the head; the caudal fin is vertical to the body axis, whereas it is horizontal to the body axis in whales and porpoises.

Fossil Sharks, Skates, and Rays

The earliest fossil sharks appeared during the Devonian Period (400 million years ago) and were called Cladodonts because their teeth were characterized by one large conical cusp with two or more lateral cusps. The 2-

meter *Cladoselache* of the late Devonian Period is the best known of the early fossil sharks. It had a terminal mouth, a large tail, two dorsal fins preceded by spines, pectoral and pelvic fins with extended projections, no anal fin, and keels on the large tail. It is believed that Cladodonts were the basal stem from which sharks evolved. *Cladoselache* disappeared by the Carboniferous Period (290 million years ago). Cladodont descendants were Hybobonts that disappeared by the late Cretaceous (65 million years ago). Hybobonts were more flexible swimming sharks possessing an anal fin, claspers in the pelvic fins of males, and two types of teeth, sharp in front and flat in rear (Springer and Gold 1989). Hybobonts provided the stock from which modern sharks evolved 190–135 million years ago. *Paleospinax*, an early representative of modern sharks, possessed better muscle attachment permitting improved swimming abilities, protrusible jaws in a ventral position, and calcified vertebrae. They and other Hybobonts disappeared by the end of the Cretaceous Period (135–165 million years ago).

Fossil great white sharks appeared in the Paleocene (65 million years ago). Whether the fossil *Cretolamna* was an earlier ancestor of the great white shark that evolutionarily led to two lines of fossil great white sharks, a large tooth line that led through *C. augustidens* and *C. subauriculatus* to *C. megalodon* and a small tooth line represented by *C. auriculatus* and *C. nodai* to modern *C. carcharias*, remains unresolved (Purdy 1996; Purdy et al. 2001). All modern sharks were developed by the Cretaceous (65 million years ago) and have changed little since.

Rays appeared in the fossil record during the upper Jurassic Period as guitarlike fishes that even had spines. All skates and rays are apparently derived from those fishes, but the pattern is unclear; however, sawfish and torpedo rays were earlier derivations from some more primitive fossil batoid group. All batoids are derived and related to Squalomorph or Squatinomorph sharks (Compagno 1977). Skeletons of fossil skates and rays are rare; instead, teeth or vertebrae are the common remnants of this vast group of elasmobranchs. Teeth of *Aetobatus*, *Rhinoptera*, and *Dasyatis*, as well as their tail spines, are common in the fossil Miocene deposits of the Carolinas.

General Systematics and Survey Studies of Western Atlantic Sharks, Skates, and Rays

Garman's (1913) monumental work on sharks, skates, and rays set the stage for later thorough studies of modern sharks by Bigelow and Schroeder (1940, 1948, 1957) and Compagno (1973, 1977, 1984, 1991, 1999). These were followed by excellent studies on various shark species or families: ocean whitetip sharks (*Carcharhinus longimanus*) (Backus, Springer, and Arnold 1956); the genus *Carcharhinus* (Garrick 1967a, 1982; Garrick and Schultz 1963; S. Springer 1950b, 1963); squalidae (Bigelow and Schroeder 1957; S. Springer 1959); hammerhead sharks (C. R. Gilbert 1967a, 1967b; Naylor 1992); cat sharks (S. Springer 1966, 1979; Springer and Sadowsky 1970); the genus *Isurus* (Garrick 1967b); the genera *Scoliodon, Loxodon,* and *Rhizoprionodon* (V. G. Springer 1964); marbled cat shark (*Galeus arae*) (Bullis 1967); *Mustelus* (Heemstra 1997); *Apristurus* (Nakaya 1991; Nakaya and Sato 1997); Squaloid sharks (Shirai 1992a, 1992b); dogfish sharks (Myakov and Kondyurin 1986); and lamniforms (Naylor et al. 1997). Additional information on the biology of various sharks can be found in Castro (1993a, 1993b, 1996a, 1996b, 2000); Parsons (1985); P. W. Gilbert (1963); and Springer and Garrick (1964). Skates and rays were studied by McEachran and Dunn (1998) and McEachran and Miyake (1986, 1990a, 1990b) (skates); McEachran and Martin (1978) (*Raja* sp.); and Nishida (1990) (*Myliobatis*).

Studies Dealing with North and South Carolina Elasmobranchs

The names Bell, Brimley, Coles, Gudger, Radcliffe, Smith, Wilson, and Yarrow dominate the shark literature of North Carolina. Much of their researches dealt with the sharks and rays found near Beaufort or in Cape Lookout bight (Gudger 1913a, 1913b). Their reports were laced with vivid experiences while harpooning or catching sharks (Schwartz 1989b). Their efforts were associated with the Ocean Leather Corporation shark fishery, which flourished between 1919 and 1922.

Coles, in 6 papers (1910, 1913, 1915, 1916, 1919, 1926), reported that 16 species of sharks were collected off Cape Lookout (Carteret County). Brimley (1935a, 1935b) noted the earliest occurrence of basking (*Cetorhinus maxi-*

mus) and whale (*Rhincodon typus*) sharks in North Carolina. Radcliffe (1913, 1916) noted 20 species of sharks frequenting Beaufort in 1913–14. Gudger was the most prolific student of sharks of the Beaufort area, publishing 11 papers (1907, 1910, 1912, 1913a, 1913b, 1915, 1932, 1947, 1948a, 1948b, 1949). Schwartz (1984, 1989a, 1989b) reported 50 species of sharks from North Carolina. Since then 6 species of sharks (*Echinorhinus brucus, Cirrhigaleus asper, Odontaspis ferox, Carcharhinus altimus, C. galapagensis, and C. perezi*) and one skate (*Neoraja carolinensis*) have been added to the North Carolina elasmobranch fauna (McEachran and Stehmann 1984; Schwartz 1993b, 1998, 2000a, 2000b; Schwartz, Jensen, and Hopkins 1995; Sheehan 1998).

Few studies have dealt with the elasmobranchs of South Carolina. Jordan and Gilbert (1882) listed 14 species in 1882. Burton (1936) expanded the list to 22 species. Bigelow and Schroeder (1948, 1953) recorded 30 species of sharks from South Carolina, Bearden (1961, 1965a) later reported 47 species, and Moore and Farmer (1981) noted 31 species of elasmobranchs from the state.

Characteristics

Teeth

Shark teeth are composed of a dentine pulp core surrounded by a layer of enamel and are embedded in connective cartilage tissue of the jaw (Kemp 1999). Teeth of living sharks are white, whereas fossil teeth are gray, blackish, or sometimes pink (Maysville quarry, N.C., personal observation). Replacement of a lost tooth or tooth row is by new teeth growing from inside and moving forward toward the mouth (Kemp 1999). Replacement rate, singly or a complete row, varies from 8 to 10 days (*Negaprion*) to 5 weeks (*Squalus canicula*) or more (Luer, Blum, and Gilbert 1990; Moss 1967). Tooth size varies with body length (Moss 1967). Each jaw possesses different types of teeth depending on their use and the shark's diet. Beginning at the midpoint of each jaw and progressing laterally and posteriorly, there are varying numbers of symphysial, anterior, intermediate, lateral, and posterior teeth (Compagno 1988). These teeth can be used for different types of feeding: cutting (tiger shark), grasping (sand tiger shark), clutching (nurse shark), vestigial (whale shark), crushing (dusky smooth hound), cutting-grasping (*Hemipristis*), grasping-cutting (mako shark), grasping-crushing (sand tiger shark), and clutching-crushing (*Heterodontus*) (Kent 1994; Kemp 1999; Frazetta 1988). Shark teeth can be serrated, smooth edged, wide, pointed, with or without cusplets at the base, and on occasion a shark will be found where "normal" shaped teeth are reversed in shape (blue shark) (Litvinov 1982, 1983). The shape of the teeth in the upper jaw may be different from those in the lower jaw. Sexual dimorphism is evident in Atlantic sharpnose shark teeth (V. G. Springer 1964) and Atlantic stingray teeth (Kajura and Tricas 1996). Each shark, skate, or ray has a distinct tooth count in each jaw. For example, a jaw count of 14-2-14/14-1-14 in a dusky shark means there are 2 symphysial teeth at the midline and 14 teeth on each side of the upper jaw, whereas 14-1-14 means there is only one symphysial at the midpoint of the jaw and 14 teeth laterally on each side of the lower jaw.

The largest shark teeth belong to the living great white shark (*Carcharodon carcharias*) and fossil great white sharks (*Carcharodon, Carcharocles = Procarcharodon* of earlier designation). The largest living great white tooth measured 6.8 centimeters high (Randall 1973); the Miocene fossil great white shark largest tooth was 18 centimeters high and 13 centimeters wide (Kent 1994). Other large fossil great white shark species teeth measure 13 centimeters (*Carcharodon chubutensis*) (Miocene Age) and 17 centimeters (*Carcharodon megalodon*) (Miocene to Pliocene age) (Kent 1994). Fossil shark teeth have been used to track phyletic changes through time (Naylor and Marcus 1994). Fossil great white shark teeth can be found in North Carolina at Aurora (Lee Creek Mine), Maysville Quarry, New Bern, Castle Hayne, Snows Cut (New Hanover County), Fussell Quarry (Duplin County), and Pungo River (Carter et al. 1988; personal observations). Fossil teeth are known from Charleston, South Carolina. Other fossil shark teeth known from the Carolinas are *Hexanchus, Carcharocles, Echinorhinus, Isurus (=Oxyrhina), Cretolamna,* and *Sphyrna* (Kent 1994, 1999a). See Case (1982) and Purdy et al. (2001) for more fossil teeth descriptions.

Sawfishes (batoids) have an elongate rostrum snout armed with varying numbers of sharp lateral teeth (see species description). Thorson (1973) dispelled the belief that there was a sexual dimorphism in the number of rostral teeth. Also, once a tooth is lost, it is not replaced (Slaughter and

Springer 1968). The rostrum is an elongation of a undifferentiated mesenchymal bud of cells located in the snout, when the embryo is about 32 millimeters long. Rostral teeth begin to form when the embryo is about 104 millimeters total length in *Pristis pectinata* (smalltooth sawfish) (Miller 1995). The smalltooth sawfish possesses 25 to 32 rostral teeth (Adams and Wilson 1995). A membrane covers the teeth and rostrum when the embryo is born, but it is quickly torn as the fish swings the rostrum back and forth impaling its food.

The last large sawfish caught in North Carolina was in June 1963 (Schwartz 1984). Recently a 1.2-meter smalltooth sawfish was gill netted and released in the "Drain" of Barden's Inlet (Carteret County) on 15 July 1999 (personal observation). Smalltooth sawfish rostrums vary in length: 405-millimeter-long rostrums have 27 right/28 left teeth; 728 millimeters long, 25 right/25 left teeth; 808 millimeters long, 26 right/24 left teeth; 830 millimeters long, 27 right and left teeth (personal observation).

Skates and rays have small pointed or flat pavementlike teeth. Pointed teeth are used during feeding or holding onto a female while mating. Flat pavement teeth found in stingrays are used to crush mollusks, crustaceans, and other hard-shelled foods. Fossil platelike teeth of *Aetobatus* and *Myliobatis* are also common in the Carolinas (Kent 1999b). Feduccia and Slaughter (1994) believed skate teeth exhibited sexual dimorphism, but that was dispelled by McEachran (1977a).

Vertebrae

Large sharks and sharklike rays tend to possess more vertebrae than do smaller sharks. Springer and Garrick's (1964) extensive study of shark vertebrae found that sharks may have 60 to 419 vertebrae; Compagno (1999) expanded that range to 477. Shark vertebrae are circular calcified disks possessing cone depressions at each end.

Vertebrae extend from the skull (neurocranium) to the end of the tail. The vertebral column consists of monospondylous precaudal vertebrae extending from the skull to the pelvic fin area and diplospondylous vertebrae prior to and onto the caudal fin. One monospondylous vertebra occupies the area between one myomere or muscle of the body, while two diplospondylous vertebrae occur within each myomere. Monospondylous vertebrae are usually larger (longer and higher) than diplospondylous vertebrae. The transition area between mono- to diplospondylous vertebrae usually occurs near the base of the pelvic fins. Diplospondylous vertebrae decrease in size as they end in the tail.

Batoids, skates, and rays possess mono- and diplospondylous vertebrae; however, the vertebral column configuration varies between species. A fused tube of vertebrae, often involving 8 to 45 vertebrae, usually extends posteriorly from the chondocranium. Varying numbers of mono- and diplospondylous vertebrae occur depending on the type of tail, which is stout in skates but short or long and whiplike in stingrays. No comprehensive analysis of the number of vertebrae exists for skates or rays, but cownose rays possess over 100 total vertebrae (personal observations).

Fin and Tail Spines (Sting)

Ten species of sharks frequenting Carolinian waters possess grooved or ungrooved dorsal fin spines: arrowhead dogfish, gulper shark, black dogfish, lined lantern shark, broadband lantern shark, blackbelly dogfish, roughskin spurdog, piked (spiny) dogfish, Cuban dogfish, and shortspine dogfish. Injuries or lacerations can occur while handling these sharks. Only the piked (spiny) dogfish spines are considered venomous. Maisey (1979) studied the development of spiny dogfish spines. McFarlane and Beamish (1987) confirmed that the spines of the spiny dogfish can be used to determine their age, which may be 70 or more years.

Skates do not possess venomous tail spines (stings). All stingrays of the Carolinas, except the smooth butterfly and some devil rays, possess venomous tail spines. The term "stings" refers to the vasodentinal spine and its developing integumentary sheath and venom glands. The term "spine" refers only to the vasodentinal portion of the sting (Halstead 1970).

Spines are found on the dorsal surface of stingray tails and are usually shed annually (personal observations), unless torn loose while fending off predators such as sharks (Bigelow and Schroeder 1953; Gudger 1932). New spines develop, depending on species, before or behind the original spine even before it is shed. Most stingrays possess 1 or 2 bilaterally serrated tail spines. Spotted eagle rays often have up to 8 spines on their tails.

Stingray spines are classified or associated with four types of stingrays (Halstead 1970):

Gymnotid, as in butterfly rays; spines are short and seldom inflict serious injuries;

Myliobatid, found in spotted eagle rays, stingrays, cownose rays, and some devil rays; the spines vary in length and can cause serious injury;

Dasyatid, found in stingrays with long tails; spines can cause great injury;

Urolophid (Urobatid), round stingrays; tail is short and spine can cause serious injury.

Tail spines are usually long, are bilaterally serrate along the length of the spine, and taper to a point. Spines are composed of an inner core of vasodentine covered by a thin layer of enamel (Halstead 1970) and possess ventral grooves separated by a ridge. Each ventral groove is lined with epidermal cells (goblet) that contain toxic fluids that are ruptured as the spine is inserted, like a hypodermic needle, into a leg, arm, or body. The resulting pain from the injury is most excruciating and, if unattended, lasts for hours. Treatment is to plunge the affected area into hot water (personal observation), which will ease the pain. If an injury is left unattended, swelling, cold sweats, and muscle spasms result that can cause death. Pulling spine(s) from a hand or foot will cause tearing of the tissues and muscles. It is better to push the spine through the affected area or have it removed surgically. Medical treatment should be sought as soon as possible at a hospital or physician's office.

Seasonality of Occurrence

All of the most common 28 species of sharks and 11 species of skates and rays that frequent Carolina estuarine, coastal, or shelf waters exhibit seasonal and depth preference patterns (Schwartz 1995, 2000a, 2000b). Many inhabit the area for only portions of the year. Bull sharks (*Carcharhinus leucas*) occur from spring to fall in coastal, inshore, estuarine, and often fresh water before migrating southward. Basking sharks move southward from cold northern waters in November and remain until waters warm up in April, causing them to migrate northward (see size section for further details). Other species simply move inshore or offshore (scalloped hammerhead, *Sphyrna lewini*) or north and south (Atlantic sharpnose shark, *Rhizoprionodon terraenovae*, and most sharks, skates, and rays) along the coast. Other species remain in deeper waters year-round (cat and lantern sharks) or inhabit deep slope or deeper ocean waters year-round (blue, oceanic whitetip, pelagic stingrays). All occurrence patterns are either linked to water temperatures, spawning movement, or feeding needs.

A typical seasonal panorama of sharks, skates, and rays in the Carolinas is as follows: From November through April, basking, spiny, and smooth dogfish sharks arrive from the north; most sharks, skates, and rays have moved offshore or south. Spiny dogfish, basking shark, and smooth dogfish migrate north in April as waters warm. Dusky, silky, blacknose, and hammerhead sharks begin to appear, as do the thornytail stingray and clearnose skate, followed by more skates and rays. From June through September, most species of sharks—bull, blacktip, blacknose, and spinner—arrive. Finetooth sharks arrive in July. From September through October, waters begin to cool, and spinner, blacknose, blacktip, hammerhead, and other sharks, skates, and rays depart offshore or move to the south. The entire cycle can be speeded up or delayed if ocean and estuarine waters are seasonally warmer or cooler than usual.

Depth Distribution

Sharks are usually found in warm temperate and tropical seas, yet they also occur in Arctic and Antarctic waters (Arctic: basking shark and porbeagle [*Lamna nasus*]; Antarctic: Ross sea to 1,232-meter depths, Antarctic starry skate, *Amblyraja georgiana*) (Bigelow and Schroeder 1965; S. Springer 1971).

Most of the 227 species of skates are found in shallow sound, estuary, inshore, or worldwide ocean waters (McEachran and Miyake 1990a, 1990b; Weitzman 1997), except western Pacific Ocean waters (McEachran and Fechhelm 1998). *Bathyraja abyssicola* (deep-sea skate), a Pacific Ocean species, is known to inhabit waters of 2,800+ meters; *Amblyraja* and other skates reach 2,300-meter depths (McEachran and Miyake 1990b).

Locally *Benthobatus marcida* (deep-sea blind electric ray), lesser electric ray (*Narcine brasiliensis*), and Atlantic torpedo (*Torpedo nobiliana*) are known in shallow to 530-meter water depths. Clearnose skates (*Raja eglanteria*) enter estuarine waters, season dependent, and ocean waters to the edge of the continental shelf (Schwartz 1995, 1996a, 1996b).

Stingrays frequent estuaries, inshore, and offshore ocean waters to 1,102-meter depths (McEachran and Fechhelm 1998). *Leucoraja garmani* (rosette skate) prefers the continental shelf edge. Seven species of stingrays frequent estuarine or low saline waters of one part per thousand salinity (Schwartz 1995, 2000a): southern stingray (*Dasyatis americana*), roughtail stingray (*D. centroura*), *D. say* (bluntnose stingray), smooth butterfly ray (*Gymnura micrura*), spotted eagle ray (*Aetobatus narinari*), bullnose ray (*Myliobatis freminvillii*), and *Rhinoptera bonasus* (cownose ray).

Size

Sharks found in Carolinian waters vary in length from a few centimeters to 12.2 meters. A whale shark (*Rhincodon typus*), the largest recorded living shark, estimated to be 12.2 meters long, was stranded at the quarantine station at the mouth of the Cape Fear River, near Southport, North Carolina, 6 June 1934 (Brimley 1935b). Several 5-meter individuals have been seen in the ocean since at the Big Rock, at buoy 13, and at Beaufort Inlet, North Carolina, in 1997, February 1999, and August 2000, respectively (J. Hiatt, personal communication, 20 August 2000).

Filter-feeding 10.2-meter-long basking sharks (*Cetorhinus maximus*), the second largest shark species after the great white shark, migrate south in winter from Arctic waters and frequent Carolinian ocean waters from November to April. Over 360, 1.1- to 7.8-meter-long basking sharks have been recorded from North Carolina waters during 1901–2002 (Schwartz 2002). Usually 1 or 2 individuals are sighted while feeding at the surface on plankton. However, a school of 15 individuals was seen 29 January 1985 off Cape Hatteras; 25 to 40 individuals 5 to 10 meters long were spotted off Cape Lookout, North Carolina, in March 1996, while 50 to 100 similar-sized basking sharks were seen as a school 1 kilometer wide, 1.6 to 3.2 kilometers long, 32 kilometers offshore between Carolina Beach and Frying Pan light, North Carolina, 11 February 1999 (M. Berry, personal communication, 11 February 1999). Large basking sharks have entered inlets and Pamlico

Sound (Schwartz 1978), and waves often roll basking sharks around in the surf and frighten winter surfboarders. They occur in North and South Carolina until ocean waters warm in spring. Thirty 6- to 9-meter basking sharks died mysteriously off Ocracoke and from Core Banks to Wrightsville Beach, North Carolina, during the cold winters of 1987 and 1988. Red tide algal bloom that occurred in the ocean at the same time did not cause their deaths.

Tiger sharks (*Galeocerdo cuvier*), another large shark, usually occur in shallow coastal waters as well as the open ocean off the Carolinas, yet they may enter sounds, inlets, and large rivers in South Carolina (Moore and Farmer 1981). The largest tiger shark (4.5 meters long, 517.7 kilograms) was captured at North Yaupon Beach, North Carolina (Coles 1919). An 807-kilogram specimen was captured 19 June 1964 at Cherry Grove pier, South Carolina (International Game Fish Association 1998). Other 3.9-meter tiger sharks were captured off Beaufort Inlet in 1993 and Shalotte, North Carolina, in 1994 (Schwartz 1998).

A 6.4-meter-long Greenland shark (*Somniosus microcephalus*), a northern species that usually inhabits cold water, was captured near Cape Hatteras in 703-meter waters (Schwartz 1989b). The only other southern sighting of a Greenland shark was made from a submersible; the shark was swimming in 2,200-meter waters off Georgia (Herdendorf and Berra 1995).

Great white sharks (*Carcharodon carcharias*) that can attain sizes of 6.4 meters (Randall 1973; Mollet et al. 1996) frequent the edge of the Carolinian continental shelf in cold ocean waters all year. A 4.8-meter, 972-kilogram specimen was captured just off the continental shelf southwest of Beaufort, North Carolina, by commercial longline fishermen 27 April 1984 (personal observation). Conversely, two 4.5-meter-long great white sharks were sighted east of Cape Lookout shoals at nearly the same place by scuba divers swimming in 36-meter-deep waters in August 2000 and July 2001. Small great white sharks (1.5 to 3 meters long) frequent cool inshore ocean or inlet waters in April (personal observations) off Beaufort and Hatteras, North Carolina.

A 3.1-meter sixgill shark (*Hexanchus griseus*) was captured at the Currituck Inlet lifesaving station, North Carolina, in 1886 (Smith 1907). Since then, several others have been captured by commercial longline fishermen off Southport, North Carolina, 24 July 1991 (C. Jensen, personal communication, 24 July 1991). Several 46- to 56-centimeter-long sevengill sharks (*Notorhynchus cepedianum*) were captured east of Southport, North Carolina, 25 March 1981, in ocean waters 89 meters deep and in 1986 in ocean waters 86 meters deep.

Scalloped hammerhead sharks (*Sphyrna lewini*) are the common 1- to 4-meter hammerhead sharks caught off the Carolinas from June to October. Seasonally, two populations of scalloped hammerhead sharks occur off North Carolina: one migrates inshore and offshore from the Gulf Stream from May to September; the other migrates north and south along the coast from June to October before moving south in the fall. Great hammerheads (*S. mokarran*) are rare June inhabitants of North Carolina ocean waters and occur seasonally during the summer in South Carolina.

Small, usually gray-colored sharks (1 to 2 meters total length), locally called "sand sharks," dominate inshore and ocean shelf waters less than 30 meters deep. Examples are blacknose shark (*Carcharhinus acronotus*), spinner shark (*C. brevipinna*), finetooth shark (*C. isodon*), blacktip shark (*C. limbatus*), sandbar shark (*C. plumbeus*), and Atlantic sharpnose shark (*Rhizoprionodon terraenovae*). All also enter rivers and sounds (Schwartz 1995, 2000a). Small sharks (0.2 to 1.5 meters long), such as *Etmopterus* sp. (0.2 meters), *Deania* (0.3 meters), *Centrophorus* (0.3 meters), and *Scyliorhinus* (1.5 meters), are usually found year-round in deep ocean waters off the continental shelf.

Skates, usually found in continental shelf waters, rarely exceed 780 millimeters in total length. Examples are clearnose skate (*Raja eglanteria*) and rosette skate (*Leucoraja garmani*). Most skates frequent sound, estuary, inshore, and offshore waters of the Carolinas during summer months (Schwartz 2000a).

Stingrays vary in size from 35 centimeters to 4.5 meters DW. Examples are bluntnose stingray (*Dasyatis say*, 30-centimeter DW); spiny butterfly ray (*Gymnura altavela*, 2-meter DW); bullnose ray (*Myliobatis freminvillii*, 2.2-meter DW); spotted eagle ray (*Aetobatus narinari*, 2.3-meter DW); and roughtail stingray (*Dasyatis centroura*, 2.1-meter DW). Most stingrays are found on or in the substrate. Several species (7) enter estuaries, sounds, and rivers (Schwartz 1995, 2000a). Spotted eagle rays can be collected in

Beaufort Inlet and the open ocean over the continental shelf. The 0.8-meter DW pelagic stingray (*Pteroplatytrygon violacea*) prefers open continental deep shelf waters. Large mantas frequent Carolinian shelf waters from July to September, yet 2-meter DW *Manta birostris* often enter sounds in late summer (Schwartz 1995, 2000b).

Color

Although most sharks are gray dorsally and white ventrally, coloration may be brown, blue, yellow, bronze, or black. Nakaya (1973) and Taniuchi and Yamagisawa (1987) reviewed the world occurrences of the 9 known cases of albinism in elasmobranchs. An albino great white shark is known from South Africa (Smale and Heemstra 1997). Recently a white tiger shark (*Galeocerdo cuvier*) was collected in the Gulf of Mexico (Rider, Athorn, and Bailey 2002). White smooth dogfish, basking, and porbeagle sharks are known from Norway and the Faroe Islands (Taniuchi and Yamagisawa 1987; Frøiland 1975). The only albino sharks and rays known from the Carolinas are an albino scalloped hammerhead shark captured September 1969 off Doboy Sound, South Carolina (McKenzie 1970) and a white southern stingray (*D. americana*) from Pamlico Sound (near Stumpy Point) captured October 1975 (Schwartz and Safrit 1977). White specimens of cownose rays, which frequent all waters of the Carolinas, are known from nearby Tangier Sound, Maryland, and lower Chesapeake Bay (Schwartz 1959; Joseph 1961). Ishihara, Honma, and Nakamura (2001) noted albino manta rays in the western Pacific.

A number of sharks have a bronze-cast body coloration, especially in late spring and early fall (blacknose, finetooth, and blacktip sharks). Finetooth, blacktip, bignose, and spinner sharks usually have a distinct whitish "Z" located on their sides beginning at the pelvic fins and projecting forward toward the pectoral fins. The posterior margins of the dorsal, pectoral, and caudal fins of the roughskin spurdog are white. A white patch occurs at the rear base of the first dorsal fin of porbeagle sharks. White spots are scattered over the body of the Atlantic sharpnose and spiny dogfish sharks. Tips of the first dorsal and pectoral fins of the oceanic whitetip shark are white. A black blotch area occupies the axils of the pectoral fins of the great white shark. Gray bars or stripes occur on whale, blotched cat, and roughtail cat sharks. Grayish-black reticulations occur on the tan-colored chain cat shark body. Gray spots are scattered on the sides of the sand tiger shark. Small blue or black spots may occur on the body of young tan-colored nurse or blacknose sharks captured in summer or fall. Lemon sharks (*Negaprion brevirostris*) are yellowish-green to gray colored, depending on time of year. Blacknose sharks (*C. acronotus*) may be yellowish in spring and usually summer, when they are incorrectly identified as lemon sharks. Blue sharks (*Prionace glauca*) are usually deep blue to black dorsally.

Skates are usually tan- to chocolate-colored dorsally. The Carolina skate (*Neoraja carolinensis*) is the only local skate with a black ventral abdominal surface (McEachran and Stehmann 1984). Spotted eagle rays possess white spots on a chocolate or blue-black background. Fall specimens of the southern stingray and cownose rays, usually uniformly brown, may have white or gray spots on the dorsal surface when stressed (personal observation). Rosette skates have rosette markings dorsally, while the clearnose skate has stripes or bars on the dorsal surface. The bluntnose stingray disc edges are orangish-yellow, as are the sides of the tail.

Deformities

Spinal and cranial deformities have been reported for bull, sandbar (*C. plumbeus*), spiny dogfish (Hoenig and Walsh 1983; Schwartz 1973), and dusky sharks (1,200-millimeter fork length, 16 June 1975, personal observation). Wing deformities in skates and rays are usually the result of bites by sharks or other individuals during mating or feeding activities.

Reproduction

Elasmobranchs exhibit several types of reproduction: oviparity and various types of viviparity (Hamlett 1999; Luer and Gilbert 1991). Egg cases are developed in 40 percent of all sharks, but not stingrays (oviviparity). Skates usually attach egg cases (mermaid purses) to substrates or vegetation for external development. The whale shark can produce up to 300 egg cases that develop/hatch internally prior to birth of the living embryos (Joung et al. 1996). Shortfin mako (*Isurus oxyrinchus*) and sand tiger sharks

(*Carcharias taurus*) develop an embryo that feeds (oophagy) on other eggs and developing embryos (intrauterine cannibalism), producing one embryo in each uterus (Hamlett 1999).

Three forms of aplacental viviparity (often called ovoviviparity) occur in sharks and rays: (1) aplacental yolk-sac viviparity, where an embryo develops internally while feeding from a large yolk sac (smooth dogfish, *Squalus acanthias*); (2) aplacental development, where embryos developing within the uterus have uterine villi or trophonemata for nutrient provision and respiration (cownose ray); and (3) placental viviparity, in 10 percent of the sharks, where a placenta and uterine development occurs (blacknose shark, etc.) (Hamlett and Koob 1999).

Pupping and Nursery Areas

While pupping and nursery areas are usually elsewhere, Schwartz (1989a, 1989b) reported that six species of sharks pup their young in Carolinian waters during warm summer months. They are smooth dogfish, spiny dogfish, blacknose, sharpnose, tiger, and dusky sharks. C. F. Jensen (1998), Pratt et al. (1998), and Jensen and Hopkins (2001) recently added spinner, sandbar, finetooth, and common thrasher sharks to the list. Three great white shark pups were found dead in Ocracoke Inlet 3 May 1996 (personal observation).

The Cape Hatteras, North Carolina, and Bull's Bay and Charleston Harbor, South Carolina, areas serve as nursery areas for many of the above shark species (Schwartz 1984; Castro 1993b). Egg cases of the clearnose skate (*Raja eglanteria*) are common in the Carolinas in spring months (personal observation). Occasional near-term developed young have been aborted during the summer by the southern or cownose stingrays in North Carolina, apparently a stress response to having been captured.

Parasites

External leeches (*Stilarobdella macrotheca*) have been found on sandbar, Atlantic sharpnose, and scalloped hammerhead sharks captured in the Carolinas (Schwartz 1993a). The monogenetic trematode *Benedenella posterocolpa* has been found in the ventral surface of cownose rays (Schwartz (1993a). The leech drops off the sharks once they enter 30-parts-per-thousand saline ocean waters. The trematode remains with the cownose ray until salinities of 28 parts per thousand are encountered in estuaries (Schwartz 1993b). Argulids are other external parasites of almost all sharks frequenting the Carolinas during early fall months (personal observations).

The parasitic eel *Semenchelys parasitica* has been found in the heart of a mako shark (Benz and Caira 1995). Stingray spines have been found penetrating the heart of a dusky shark (J. Bonaventura, personal observation, 1984). Various stingray species' spines, often 50 or more, have been found in the mouth and oral cavities of hammerhead sharks (Gudger 1907, 1932). A spine of the southern stingray was retrieved from the abdomen of a bottlenose dolphin in North Carolina (McLellan, Thayer, and Pabst 1996).

Shark Attacks and Dangerous Sharks

The image of a triangular fin slicing through the water—enhanced by the movie *Jaws*—causes swimmers or boaters to panic. Shark attacks are relatively rare; however, those sharks that do attack are rarely seen by the victim. Especially dangerous are the tiger, bull, mako, dusky, blue, blacktip, great white, and hammerhead sharks. Whale and basking sharks, despite their docile habits and size, possess small needlelike or hooked teeth and should be avoided because of their overall size. Likewise, their reaction response time is very fast, and a simple flip of their tail can cause serious injury.

North Carolina has had 23 authenticated shark attacks (Table 2) between 1870 and 2002. Most were caused by bull sharks, although 3 of the 5 attacks in 2000 were caused by blacktip sharks. What causes shark attacks? Although not proven, temperature may play a role, with more attacks occurring when ocean waters are warmer than 27°C. Color of a swimsuit or scuba tank may be another factor, as attacks are usually associated with such light colors as yellow, orange, bright green, or light blue being worn at the time of the attack. Water turbidity along the shore has been associated with shark attacks as well as the fact that sharks were feeding and the victim was simply in the way. Recent comments point to or blame the increased use of the ocean (the shark's domain) by the burgeoning human population as a causal agent, so that there are more chances for an encounter.

Analyzing the shark attacks that have occurred in Pamlico Sound or the ocean off North Carolina from 1870 through 2002 reveals a remarkable correlation of factors when each attack occurred. Twenty of the recorded attacks occurred in the afternoon during ebbing tides, regardless of whether the attack took place in the ocean or in Pamlico Sound. The 2 attacks in 2001 happened during the end of a flooding tide; coastal winds made the area behave as if it were experiencing an ebbing tide. Only the 2 May 2002 attack occurred at 11:00 A.M., right at the end of the ebb tide. Likewise, each of 7 attacks occurred during full or new moons in July and August. Thus the public is advised to be extremely careful while swimming or scuba diving in waters warmer than 27°C in July and August and during full or new moons and during ebbing tides. Do not swim at night or in the early morning or evening, periods when most sharks are feeding.

South Carolina experienced 23 shark attacks in the 60 years up to 1989 (Schwartz 1984). Since then, 20 more have occurred through 2001; of these, 6 were recorded in 2001 (G. Ulrich, personal observations). Myrtle Beach, Folly Beach, and Hilton Head experienced the most shark attacks in South Carolina, while Emerald Isle, Carolina Beach, and Wrightsville Beach were frequent sites of attacks in North Carolina.

Shark Fisheries and Tournaments

A commercial fishery was established by the Ocean Leather Corporation of New York in 1919 near, at that time, the old menhaden factory in Morehead City, North Carolina (now North Carolina State University building area). The fishery was operated by Captain E. Young. Originally a pier was built extending into the deep waters of Bogue Sound. A skinning and dissecting platform was constructed on the pier to shield the sharks and their by-products from the sun's heat. A curing, storage, and by-product building was also constructed. Later a shark liver oil plant was established, complete with a narrow-gauge railroad from the end of the pier to the building and processing facilities. A shark meal and fertilizer facility was added that

TABLE 2. *Verified shark attacks in North Carolina by date, moon phase, tide stage, locality, and survival status, 1870–2002*

DATE	MOON PHASE[a]	TIDE (TIME) HIGH	LOW	LOCALITY	STATUS[b]
27 July 1870	●+	0652	1331	Southport	S
1900–1905	—	—	—	Ocracoke Inlet	D
1905	—	—	—	Ocracoke	D
29 July 1905	◑	1457	2113	Davis Shore, east of Beaufort	D
21 Sept. 1935	◑-	0646	1201	New River	D
1 July 1940	◑+	1015	1649	Holden Beach	S
6 Aug. 1945	●	0815	1507	Ocracoke	D
16 July 1957	○-	1015	1649	Atlantic Beach	D
25 Sept. 1971	○	1130	1708	Emerald Isle	S
26 Aug. 1976	●	0858	1437	Emerald Isle	S
11 Aug. 1980	●	0839	1434	Ocean Isle	S
19 Aug. 1986	○	0737	1304	Masonboro Inlet	S
15 Aug. 1993	●-	1012	1604	Pamlico Sound	S
6 July 2000	◐	1114	1339	Duck-Hatteras	S
16 July 2000	○	0832	1636	West Holden Beach	S
18 July 2000	○-	0906	1631	Wrightsville Beach	S
11 Aug. 2000	◐	0636	1431	Emerald Isle	S
2 Oct. 2000	●-	1114	1800	Wrightsville Beach	S
3 Sept. 2001	○-	1944	1326	Avon	D
3 Sept. 2001	○-	1944	1326	Avon	S
4 July 2002	◑-	1449	2109	Wrightsville Beach	S
20 July 2002	◑	1616	2050	Emerald Isle	S
5 Aug. 2002	◑	1651	1005	Topsail Beach[c]	S

[a] ● = new moon, ◐ = first quarter, ○ = full moon, ◑ = third quarter
+ means attack occurred the day after the given moon stage
- means attack occurred the day before the given moon stage
[b] S = survived; D = died
[c] Occurred 11:00 A.M. at end of ebb tide

included a steam plant, chopper, grinder, steam cooker, and long heated tunnel-dryer. Total cost was $60,000, a large sum in those days.

Three 10.7-meter open-cockpit dory-type motor-driven boats, equipped with mast and derrick to hold the shark nets, with a crew of 4 per boat, were utilized in the fishery. Shark nets (183 meters by 100 meshes 65.2 meters deep) were set on the bottom about 8 kilometers offshore. A bottom line was weighted, anchored on both ends with 23-kilogram kedge anchors. The top line was fitted with 10-centimeter corks. Sharks averaging 2.1 meters long were captured at a rate of 50 to 60 per day. Hides were skinned by a good skinner in 15 minutes, fleshed, and stored for about a week prior to shipment. Fins (both pectoral, one dorsal, and caudal fin) were hung to dry. Livers were cut into pieces and cooked in 208-liter steel drums for their oil. The chopper did not work, and the quantity of carcasses was not sufficient to operate it steadily. The operation ceased in 1922. The fertilizer plant moved to Sanibel Island, Florida, where it never operated. While proving unprofitable at Morehead City, the Ocean Leather Corporation, founded by A. Ehrenheider and managed for 25 years by L. Moresi, is still in existence (Moresi 1975).

Cecil Nelson began a small shark fishery in Morehead City in 1936 and operated it for 5 years. He reported captures of 3,000 sharks, all more than 1.8 meters long, during one April–June season. He shipped the skins and livers to northern markets and sold teeth for one cent each. The remainder of the shark was used as fertilizer by the local farmers or was thrown overboard. The chimney of his rendering plant is still evident on Phillip's Island in Newport River, just north of Morehead City, Carteret County. Shark catches during 1937–39 of 95 to 264,000 kilograms can be attributed to the Nelson fishery.

Lloyd Davidson of Morehead City reestablished a commercial longline fishery for sharks in August 1983 following increased public awareness, utilization, and demand for shark meat. This action, along with the increased sport fishery for sharks, reinstated sharks as an important resource to the state. Davison stopped fishing in 1985 after regulations and efforts made catches impractical.

Shark tournaments and sport fisheries have revived recent interest in shark tournaments in North and South Carolina (the North Carolina tournament ended after 10 years) (Schwartz 1998). Historically sharks were overlooked or considered trash, nuisances, or inedible by the public. Local ordinances prohibited shark fishing at some beaches and piers. Regardless of such restrictions, shark fishing in South Carolina provided capture of the world record all-tackle tiger shark of 801 kilograms on 14 June 1964 at Cherry Grove pier (Moore and Farmer 1981; International Game Fish Association 1998). Commercial interests waxed and waned in South Carolina and were directed for small sharks, such as dogfishes, which were sold at Georgetown and Charleston as "gray fish" or "steak fish."

Most skates and rays are simply cut up and sold as bait to catch other fishes. Skates are often sold as scallops (but note that the grain of meat is horizontal in sharks and skates, while vertical in true scallops).

Future Outlook for Elasmobranchs

Elasmobranchs arose in the Devonian age and over 400 million years radiated into the many types of sharks, skates, and rays encountered today (Compagno 1973, 1999; Zangerl 1973, 1981). Today they have adapted and fill every water niche on the planet (Compagno 1990). During the intervening 65 million years after the Cretaceous, there was no evidence of a decline in elasmobranchs (Compagno 1990), yet today there are concerns for their future existence and abundance. Most people shunned sharks in the past as harmful organisms. Only their fins were used in East Asian countries to make the tasty and expensive shark fin soup. Until 1870, people were often not even aware of shark occurrences in an area (Liu, Chen, and Joung 2001). During the intervening years of 1965–95, however, the world's use of elasmobranchs increased 92 percent, when at least 200,000 to 600,000 metric tons were landed (Compagno 1990; Liu, Chen, and Joung 2001). Sharks accounted for 60 percent of the elasmobranch catches; skates and rays, 40 percent (Bonfil 1994). Today, at least 700,000 and perhaps 1.5 million tons have been landed (Stevens et al. 2000). Commercial and sport fisheries proliferated during this period (Anderson 1990). Twenty-six countries had catches exceeding 900 tons per year.

Several intensive fisheries have developed in the United States since

1980 (anonymous 1998; Castro, Woodley, and Brudek 1999; Hueter 1991, 1998; Walker 1998). These increased fishing efforts may have caused the "collapse" of many fisheries. Atlantic Ocean fisheries saw the collapse of the porbeagle (Casey, Mather, and Hoenig 1978) and basking shark populations (Parker and Stott 1965). As these fisheries increased, affected shark species have not recovered. Sims and Reid (2002) have a different interpretation of the decline of the basking shark population (an interrelationship among basking sharks and zooplankton, not overfishing, causing the decline). During the 1960–86 period a major source of catches of large sharks was the recreational fishery (Anderson 1990), followed by longlining and Japanese tuna longline fisheries. Various fishery councils were established on the east coast to study and model the shark fisheries in efforts to "manage [and] enhance the[ir] recovery" from overfishing (Hoff and Musick 1990; Walker 1998). These efforts and concerns led to implementation of management plans regulating seasons, sizes, and catch limits of various designated shark groups: offshore pelagics (10 species), large coastal (22 species), and small coastal sharks (7 species) (National Marine Fisheries Service 1993).

Why this concern over sharks and shark populations? Sharks exhibit a number of interesting characteristics: slow growth; delayed maturation; long reproductive cycles (six months in blacknose sharks [Schwartz 1989b]; two years in spiny dogfish; three years in blue sharks); usually low fecundity (tiger sharks can have 65 young [Schwartz 1996b]); and long life spans (Hoenig and Gruber 1990). Yet at which life stage—young or adult—should sharks be harvested?

It is difficult to appraise evenly the problems facing sharks, shark fisheries, and shark populations as to whether overfishing caused their supposed declines. First, biological life history information for all species found in the Carolinas and worldwide is lacking (Castro, Woodley, and Brudek 1999). Likewise, world reviews (Bonfil 1994) of shark fisheries and populations reveal that catches off one country may have declined while they were up in an adjacent country. Second, too much catch data has depended on landing figures that may be completely inaccurate, as catches are usually recorded as the port or country of landing, not when or where the catch was made. By-catch information is also lacking and may not reflect the real data of season, gear fished, target species fished or discarded, and so forth.

Such inaccuracies have led scientists and the public to believe a species or population has been overfished. Is it "hype," job security, or just inaccurate catch numbers or weights that reflect the real situation (Mrosovsky 2002; Stevens et al. 2000)? Mortality data, to further complicate matters, is lacking for elasmobranchs (Dulvy et al. 2000), yet federal and state agencies continue (using catch statistics) devising management plans in hopes of rehabilitating a species or population.

On the other hand, sensible shark fishermen will/do self-regulate their efforts if one species or population appears to decline. They shift to the next most available species, thereby permitting the recovery of the affected species over time. Likewise, overfishing a large-sized species causes an increase in the numbers of smaller species that were held in check by the larger species, which fills the void and catches (competitive release) (Dulvy et al. 2000).

While many doom-and-gloom scientists try to answer shark management questions (Hoff and Musick 1990), shark populations will continue to wax and wane. After 35 years of fishing the same areas off Morehead City, using the same gear, from April to November, I have observed that shark populations other than the dusky shark are stable or increasing, even though management plans are or were not in effect (Schwartz 1999).

Sharks have been around for millions of years and will continue to swim the changing world oceans feeding on its inhabitants. Fishermen know well that if one is at the right place at the right time, sharks will be captured. The increase in shark populations off North Carolina in 2001 cannot be linked to management plans having "enhanced" the existence of a species. Regulations cannot prove that they caused the increases, for if populations are not abundant or increasing, then why can/do fishermen fill their fishing quotas in record time during any regulated fishing season?

What will happen now that the great white, tiger, and other species of sharks are protected here and in Hawaii, South Africa, and Australia? Will we be overwhelmed with sharks and will that increase the frequency of shark attacks of the public? How will they upset the ocean ecosystems? Sensible fisheries will self-regulate levels of harvesting of shark populations, for which scientific models, as yet, provide no conclusive answers.

Meanwhile, sharks have played a role in both sport and commercial fisheries of the Carolinas off Cape Hatteras, Ocracoke, Morehead City, Beaufort, Wrightsville, and Southport, North Carolina. Fishing for sharks is discouraged in South Carolina for safety reasons by operators of ocean piers, although shark fishing is often permitted when fishing is slack. Young ("sand sharks") sharks usually plague pier fishermen from May to July (blacknose, dogfish, Atlantic sharpnose sharks), another good sign sharks continue to flourish.

Skates and rays do not contribute to economic fisheries within the Carolinas, except for some skates whose wings are used as substitute scallop meat. However, several species have quietly disappeared from various areas of the Atlantic. Most notably, *Dipturus laevis,* a species common 45 years ago, is now rare or absent (Casey and Myers 1998). It appears that body size is the best indicator of vulnerability of overfishing of skates and rays to extinction (Stevens et al. 2000), for populations of the large-sized *Raja eglanteria* are increasing while smaller-sized *Malcoraja ocellata* and *Leucoraja erinacea* populations are decreasing (Dulvy and Reynolds 2002; Dulvy et al. 2000; Stevens et al. 2000).

species accounts

sharks

Atlantic angel shark

Squatina dumeril Lesueur, 1818

DISTINGUISHING FEATURES: Only dorso-ventrally flattened diamond-shaped shark. Snout blunt. Gill slits positioned on the sides of the head. Eyes on top of the head. Large spiracles located on head just behind the eyes. Jaws protusible. Pectoral fins are separate from the head and pelvic fins. Two dorsal fins near the tail; no ridge between dorsal fins. No anal fin. Teeth 10-10/9-9.

COLOR: Brown above. Some dark blotches on dorsal surface. White ventrally.

DISTRIBUTION AND OCCURRENCE: Known from Massachusetts to Florida, Gulf of Mexico, Central America, south to Venezuela. A coastal inshore transient September to April. Large September catches in 73-meter-deep waters suggest abundance in offshore Carolinian waters during other seasons.

Usually occurs in North Carolina from Core Sound to offshore ocean waters 42 meters deep; rare in South Carolina. Elsewhere known to 1,270-meter depths. Best North Carolina catches are at Cape Hatteras (J. W. Dodril, personal communication, September 1984), Beaufort Inlet, and near Cape Fear River.

SIZE: Attains 1.2-meter lengths in North Carolina and elsewhere.

REFERENCES: Bigelow and Schroeder 1948; Clark and Kristoff 1990; Compagno 1984; Gadiz et al. 1999; Kohler, Casey, and Turner 1998; McEachran and Fechhelm 1998; Radcliffe 1916; Schwartz 1984; Shirai 1992a; Smith 1907.

Bluntnose sixgill shark

Hexanchus griseus (Bonnaterre, 1788)

DISTINGUISHING FEATURES: Large stout body possessing six gill slits on side of head. Single dorsal fin far back on body. Large blunt head, eyes small. Six rows of bladelike comb-shaped teeth on each side of lower jaw. Teeth 20-20/16-1-16.

COLOR: Pale gray dorsally, brownish ventrally. Unevenly spotted with brown spots. May have conspicuous white spot (pineal spot) on top of head.

DISTRIBUTION AND OCCURRENCE: A worldwide species in temperate and tropical waters, North Carolina, Caribbean Sea, and Gulf of Mexico to 1,875-meter water depths.

Known in North Carolina since 1886 when 3.1-meter specimen was caught at Currituck Inlet lifesaving station. Four others were caught: two 13 December 1969 and two 8 August 1987 and 26 July 1991 near Frying Pan lightship (personal observations). South Carolina observations were from a submersible 155 kilometers southeast of Charleston, depth 400 meters, 27 August 1986.

SIZE: Overall size attained 4.8 meters. Bluntnose sixgill sharks caught in North Carolina were 3.1 meters long, weighed 113 kilograms.

REFERENCES: Barans and Ulrich 1996; Bigelow and Schroeder 1948; Carey and Clark 1995; Clark and Kristoff 1990; Compagno 1984; Gilken and Coad 1989; McEachran and Fechhelm 1998; Nevell 1998; Smith 1907.

Sharpnose sevengill shark

Heptranchias perlo (Bonnaterre, 1788)

DISTINGUISHING FEATURES: A slender shark. Head narrow pointed. Big eyes. Seven gills small. Single dorsal fin far back on body. Pelvic fins reach anal fin. Five rows of bladelike, comb-shaped teeth in lower jaw. Teeth 21-23/17-1-17.

COLOR: Gray dorsally, light ventrally. No body spots. Dorsal fin and caudal fin lobes black tipped.

DISTRIBUTION AND OCCURRENCE: Tropical and temperate seas, North Carolina to West Indies on continental shelf inshore and to 1,000-meter depths.

North Carolina specimens, 46 and 56 centimeters long, were captured at 32°54′N, 77°41′N by commercial fishermen in 1986 and 1987. Waters were 248 meters deep (water temperature 10.4°C). Six other small sevengill sharks were captured off Southport in 1981, 1986, and 1987 in waters 302 meters deep. Found in stomachs of wreckfish, *Polyprion americanus*, caught in South Carolina.

SIZE: Attains 1.4-meter lengths elsewhere. Smaller (see above) in the Carolinas.

REFERENCES: Bigelow and Schroeder 1948; Clark and Kristoff 1990; Compagno 1990; Garrick and Paul 1971; Schwartz 1984.

BARBEL PRESENT ON EACH NOSTRIL

Nurse shark

Ginglymostoma cirratum (Bonnaterre, 1788)

DISTINGUISHING FEATURES: Body stout. Head large in adults, terminal mouth. Ventrally, a single barbel at front edge of each nostril. No keels on caudal peduncle; whale sharks have nostril barbels and caudal peduncle keels. First dorsal fin far back on body over pelvic fins. Two dorsal fins of nearly equal size lack spines; first dorsal fin longer than second. Spiracles, openings behind eyes, present. Five gill slits. Pectoral fins rounded in shape. Teeth 24-38/22-32.

COLOR: Tan to chocolate brown throughout, often with a yellowish sheen to the body. Young have 2- to 3-millimeter black or blue dots or spots on the body and tail. Adults have two dark spots high on the sides anterior to the dorsal fin.

DISTRIBUTION AND OCCURRENCE: Worldwide in tropical and temperate waters to 70-meter depths; Rhode Island to southern Brazil. A sporadic visitor to the Carolinas south of Cape Hatteras and Johnny Mercer Pier (New Hanover County), North Carolina. Specimens have been caught in shore or inlet waters less than 12 meters deep November 1985, August 1989, and July 1995 and 2002. Enters sounds and rivers in South Carolina.

SIZE: Sizes range to 4.3 meters total length. Largest North Carolina specimen 2.1 meters total length, 1 July 2002.

REFERENCES: Bigelow and Schroeder 1948; Castro 2000; Compagno 1984, 2001; Garrick and Schultz 1963; Jordan and Gilbert 1882; Kohler, Casey, and Turner 1998; Luer, Blum, and Gilbert 1990; Moore and Farmer 1981; Radcliffe 1916; Schwartz 1984; S. Springer 1963.

NO ANAL FIN, NO SPINES IN DORSAL FINS

Bramble shark

Echinorhinus brucus (Bonnaterre, 1788)

DISTINGUISHING FEATURES: No anal fin. No spines at front of dorsal fins. First dorsal fin located over pelvic fins. Heavy buckler-like denticles in clumps scattered over body giving warty appearance. Teeth 11-11/10-10.

COLOR: Chocolate brown throughout body; color slightly darker ventrally. Some specimens may be dark gray dorsally and whitish ventrally.

DISTRIBUTION AND OCCURRENCE: Worldwide, in the Atlantic from Cape Cod to Brazil usually in deep waters to 900 meters; occasionally moves into shallow shelf waters 27 meters deep.

A 2.2-meter specimen (third for the Atlantic) was captured 20 January 1968 off Virginia, while a 2.7-meter specimen (fourth known in the western Atlantic) was caught by longline 15 March 1982 just north of Cape Hatteras (depth 27 meters). A fifth specimen 2.1 meters long is known from Georgia (Thompson, personal communication, September 1999).

SIZE: Attains length of 3.1 meters. In North Carolina, one 2.7-meter-long specimen known.

REFERENCES: Bigelow and Schroeder 1948; Compagno 1984, 1990; Garrick 1960a; Joel and Ebenzer 1991; Musick and McEachran 1969; Pfeil 1982; Schwartz 1993b, 2000b.

Greenland shark

Somniosus microcephalus (Bloch and Schneider, 1801)

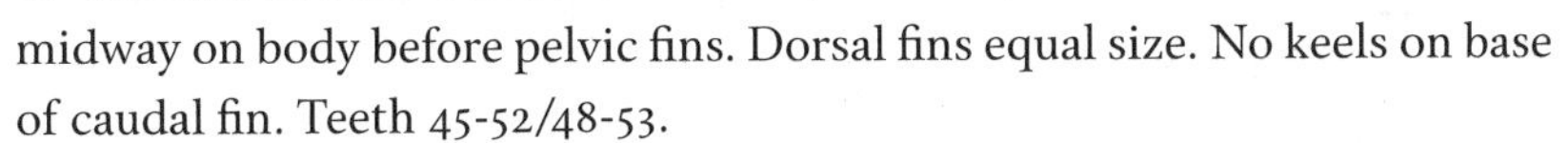

DISTINGUISHING FEATURES: A large stout shark. Short, rounded snout. No anal fin. No spines in front of dorsal fins. First dorsal fin midway on body before pelvic fins. Dorsal fins equal size. No keels on base of caudal fin. Teeth 45-52/48-53.

COLOR: Gray, blue-black, brown; may have velvet sheen to body color.

DISTRIBUTION AND OCCURRENCE: Arctic and North Atlantic to Gulf of St. Lawrence; strays south as far as Georgia in deep water, often 2,200 meters deep. Known from single specimen captured in North Carolina off Cape Hatteras.

SIZE: Attains 6.4-meter lengths. Smaller in North Carolina.

REFERENCES: Bigelow and Schroeder 1948; Compagno 1984; Garrick and Schultz 1963; Herdendorf and Berra 1995; Schwartz 1989b.

Kitefin shark

Dalatias licha (Bonnaterre, 1788)

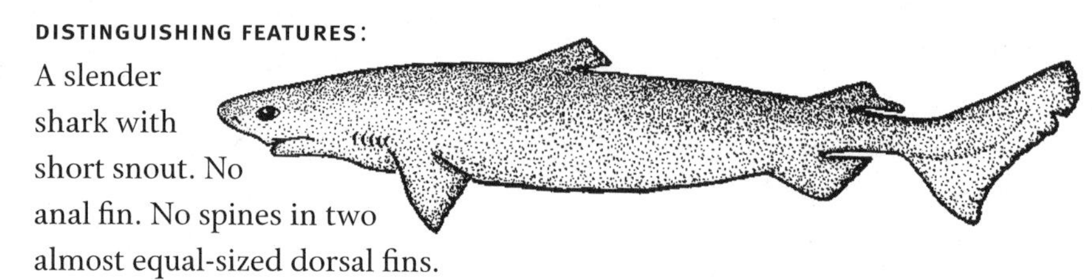

DISTINGUISHING FEATURES: A slender shark with short snout. No anal fin. No spines in two almost equal-sized dorsal fins. Snout length is less than the distance from the mouth to the pectoral fin origin. First dorsal fin originates posterior to the pectoral fins. Second dorsal fin begins midway behind the pelvic fin origin. Interdorsal fin distance greater than the distance from tip of snout to axil of pectoral fin. Teeth 8 or 9-8 or 9/8 or 9-1-8 or 9.

COLOR: Uniform dark gray-black dorsally, whitish ventrally.

DISTRIBUTION AND OCCURRENCE: Georges Bank to Gulf of Mexico in waters to 1,648 meters deep. Known from deep waters in South Carolina (S. van Sant, personal communication, December 2001).

SIZE: Maximum total length 182 centimeters.

REFERENCES: Bigelow and Schroeder 1948, 1957; Clark and Kristoff 1990; Compagno 1984; Garrick 1960b; Moore and Farmer 1981.

NO ANAL FIN, SPINES IN DORSAL FINS

Arrowhead dogfish

Deania profundorum (Smith and Radcliffe, 1912)

DISTINGUISHING FEATURES: Slender body shape. No anal fin. Both dorsal fin spines possess lateral grooves on each side. Long snout one-half length of head. Snout longer than distance from mouth to pectoral fin origin. First dorsal fin smaller than second dorsal fin. A subcaudal keel on caudal peduncle. No anal fin. Teeth 26-31/26-30.

COLOR: Gray or brown; black.

DISTRIBUTION AND OCCURRENCE: Usually found in eastern Atlantic and Pacific oceans; North Carolina in western Atlantic. A single North Carolina capture, reported as *Deania elegans*, caught 23 February 1958 at 34°40′ N, 75°32′ W, in 366-meter waters.

SIZE: About 76 centimeters total length.

REFERENCES: Bigelow and Schroeder 1957; Compagno 1984; Garrick 1960b; S. Springer 1959.

Gulper shark

Centrophorus granulosus (Bloch and Schneider, 1801)

DISTINGUISHING FEATURES: No anal fin. Large grooved spines in both dorsal fins.

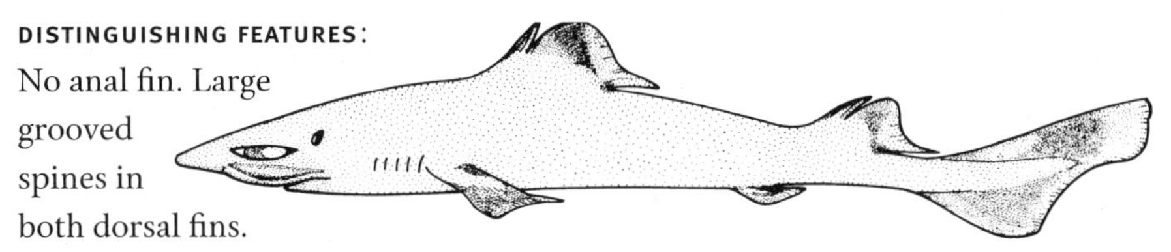

Slender, with long broad snout. Snout shorter than distance from mouth to origin of pectoral fins. First dorsal fin origin over axil of pectoral fin. Distance between dorsal fins equals distance from tip snout to axils of pectoral fins. Rear tips of pectoral fins strongly extended and reach beyond first dorsal fin spine. Teeth 31-42/27-35.

COLOR: Brown throughout.

DISTRIBUTION AND OCCURRENCE: North Carolina to Gulf of Mexico in waters to 1,200 meters deep. Rare off the Carolinas; one from North Carolina, February 1984 (USNM 22019), and one in South Carolina (S. van Sant, personal communication, December 2001).

SIZE: Maximum size 150 centimeters total length.

REFERENCES: Bigelow and Schroeder 1957; Bigelow, Schroeder, and Springer 1955; Clark and Kristoff 1990; Compagno 1984.

Black dogfish

Centroscyllium fabricii (Reinhardt, 1825)

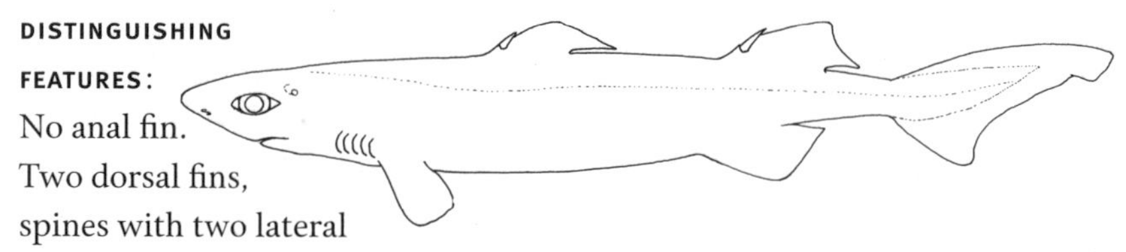

DISTINGUISHING FEATURES: No anal fin. Two dorsal fins, spines with two lateral grooves. No lateral ridge between pelvic and caudal fins. Second dorsal fin larger than first. Pectoral fins' rear edges end before first dorsal fin spine origin. Caudal peduncle short; distance from second dorsal fin insertion to upper caudal fin origin as long as distance from eye to first gill slit. Teeth 34-0-34/34-0-34.

COLOR: Brown to black throughout. No markings on body or tail. Luminescent organs scattered over skin.

DISTRIBUTION AND OCCURRENCE: Baffin Island, Canada to Virginia. No specimens from the Carolinas, but expected in deep waters near edge of continental shelf. Known to 1,600-meter water depths.

SIZE: Attains 1.1-meter total lengths.

REFERENCES: Bigelow and Schroeder 1948, 1954, 1957; Yano 1995.

Lined lantern shark

Etmopterus bullisi Bigelow and Schroeder, 1957

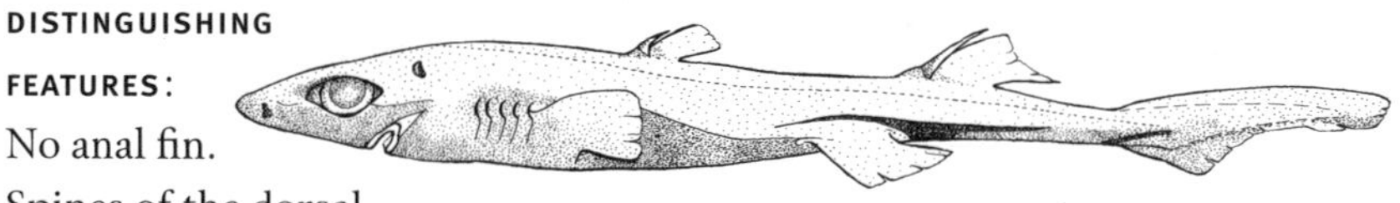

DISTINGUISHING FEATURES: No anal fin. Spines of the dorsal fins have lateral grooves. Slender; interspace between dorsal fins less than distance from tip of snout to first gill slit. Long tail. Pectoral fins reach level of first dorsal fin origin. No precaudal pit. No ridge on caudal peduncle. Teeth 18-20/27-31 on each side of jaw.

COLOR: Black all over. Elongate black flank markings above and behind pelvic fins.

DISTRIBUTION AND OCCURRENCE: Western Atlantic Ocean from North Carolina to the Caribbean and Honduras. Only North Carolina specimens collected February 1957 in 366-meter water.

SIZE: Attains 24-centimeter total lengths.

REFERENCES: Bigelow and Schroeder 1957; Compagno 1984; McEachran and Fechhelm 1998; Schwartz and Burgess 1975.

Broadband lantern shark

Etmopterus gracilispinis Krefft, 1968

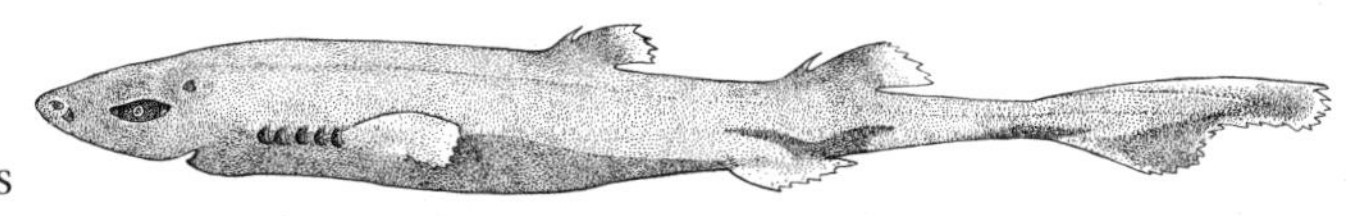

DISTINGUISHING FEATURES: No anal fin. Two dorsal fins with lateral grooved spines. Interspace between dorsal fins less than distance from snout to first gill slit. Pectoral fins end far short of first dorsal fin origin. Second dorsal fin larger than first. Ends of dorsal, pectoral, pelvic, and caudal fins may be frayed. No ridge on caudal peduncle or precaudal pit. Tail short. Teeth 18-20/27-37 on each side of jaw.

COLOR: Brown above to black below. Distinct winglike flank markings above and behind pectoral fins and at caudal base and along its axis.

DISTRIBUTION AND OCCURRENCE: North Carolina to Florida, Uruguay to Argentina, to 600-meter depths. Two North Carolina female specimens captured February 1976 (126- and 221-millimeter total lengths). Also known in South Carolina (S. van Sant, personal communication, December 2001).

SIZE: Maximum size 33 centimeters total length.

REFERENCES: Bigelow and Schroeder 1957; Compagno 1984; Krefft 1968b, 1980; McEachran and Fechhelm 1998; Schwartz and Burgess 1975; Sadowsky, Arfelli, and de Amorim 1986.

Caribbean lantern shark

Etmopterus hillianus (Poey, 1861)

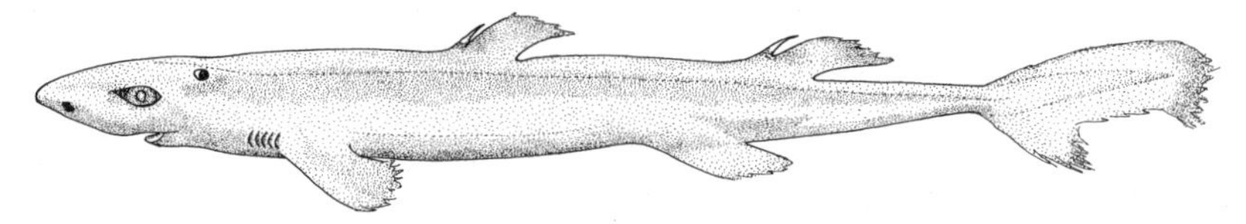

DISTINGUISHING FEATURES: No anal fin. Body slender. Two dorsal fins with lateral grooved spaces. Interspace between dorsal fins more than distance from snout to first gill slit. First dorsal fin origin located posterior to rear of pectoral fins. Pectoral fins forward of dorsal fin origin. Second dorsal fin higher than first. Edges of fins not frayed, covered with skin. Teeth 12-13/18-19.

COLOR: Gray to brown above, black below. Long broad black area above and behind pectoral fins; black area on ventral caudal peduncle, across caudal fin, and at tip of upper caudal fin lobe.

DISTRIBUTION AND OCCURRENCE: Virginia to southern Florida, northern Gulf of Mexico and some Caribbean islands in waters to 717 meters deep.

SIZE: Attains 50-centimeter lengths. North Carolina specimens attain 24.2-centimeter lengths in 310-meter waters year-round. Similar sizes occur at 300-meter depths in South Carolina.

REFERENCES: Bigelow and Schroeder 1948, 1957; Compagno 1984; Schwartz 1984, 1989b; Schwartz and Burgess 1975.

Roughskin spurdog

Cirrhigaleus asper (Merritt, 1973)

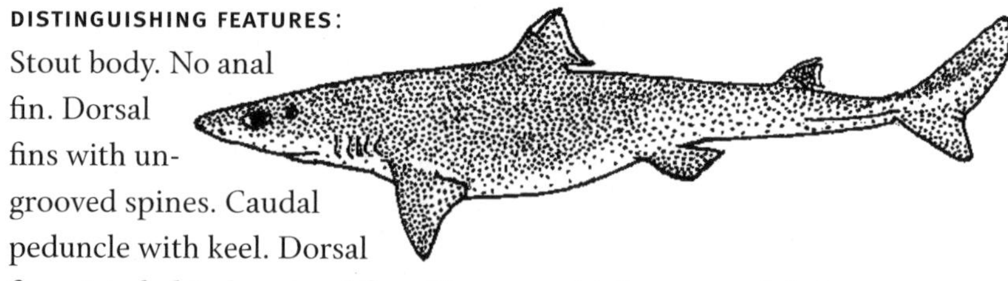

DISTINGUISHING FEATURES: Stout body. No anal fin. Dorsal fins with ungrooved spines. Caudal peduncle with keel. Dorsal fin origin behind pectoral fins. Upper caudal fin precaudal pit may or may not be present. First dorsal fin equals second dorsal fin in size. Distance between dorsal fins greater than distance from tip of snout to origin of pectoral fin. Teeth 24-28/22-27.

COLOR: Gray above, lighter below. No spots on body. Dorsal fins, pectoral and caudal fins edges white, especially the rear edge of each dorsal fin.

DISTRIBUTION AND OCCURRENCE: North Carolina to Gulf of Mexico in depths of 200 meters. North Carolina specimens captured at 32°38′ N, 74°414′ W, 18 July 1982.

SIZE: Attains 118-centimeter lengths.

REFERENCES: Compagno 1984; Garrick 1960b; McEachran and Fechhelm 1998; Merritt 1973; Schwartz 2000b.

Piked (spiny) dogfish

Squalus acanthias Linnaeus, 1758

DISTINGUISHING FEATURES: No anal fin. Slender round-bodied shark often confused with *S. cubensis* and *S. mitsukurii*. Dorsal fin spines lack lateral grooves. No interdorsal ridge. Dorsal fin positioned well behind pectoral fins. Distance from fifth gill slit to first dorsal fin spine about equal to distance from front of eye to fifth gill slit. Pectoral fins' posterior margins concave. Caudal peduncle with lateral ridge. Teeth 14-0-14/11 or 12-0-11 or 12.

COLOR: Gray above, white below. Conspicuous white spots on sides of body. No white edge to dorsal and other fins, which may be dusky tipped in young.

DISTRIBUTION AND OCCURRENCE: Greenland to Cuba and Uruguay to Argentina. Enters estuaries and rivers; inshore and well offshore in ocean waters to 900-meter depths in large or small congregations. Support winter commercial fisheries in North Carolina November to April. Some larger females remain until May or September before moving offshore or northward as water temperatures warm above 18°C (usually 12°C). Similar occurrences in South Carolina.

SIZE: Females attain sizes of 1.5 meters long; males are shorter. Pup young in April–May and comprise the spring pier fisheries of "sand sharks."

REFERENCES: Al-Badri and Lawson 1985; Bearden 1965a, 1965b; Bigelow and Schroeder 1948, 1957; Garrick 1960b, 1960c; A. C. Jensen 1966; C. F. Jensen 1998; Jensen and Hopkins 2001; Lawson 1709; Moore and Farmer 1981; Myakov and Kondyurin 1986; Pratt et al. 1998; Radcliffe 1916; Schwartz 1973, 1989b, 2000a.

Cuban dogfish

Squalus cubensis Howell-Rivero, 1936

DISTINGUISHING FEATURES:
No anal fin.
No lateral grooves in dorsal fin spines.
Long snout, long caudal fin.
Snout shorter than distance from mouth to origin of first dorsal fin. First dorsal fin origin anterior to rear corner of pectoral fin. Second dorsal fin shorter than first. Distance between dorsal fins greater than distance from tip of snout to first gill slit. Posterior margins of fins not fringed. Caudal peduncle with lateral ridge. Teeth 13-0-13/13-0-13.
COLOR: Gray dorsally, white ventrally. Dorsal fins black tipped; other fin edges white. No spots on body.

DISTRIBUTION AND OCCURRENCE: North Carolina to Gulf of Mexico, Mexico, and off southern Brazil and Argentina in waters to 380 meters deep. Occurs April to November in North Carolina waters 311–437 meters deep.

SIZE: Attains lengths of 110 centimeters.

REFERENCES: Bigelow and Schroeder 1948, 1957; Bigelow, Schroeder, and Springer 1953; Compagno 1984; Garrick 1960b; Schwartz 1989b; Sadowsky and de Amorim 1981.

Shortspine spurdog

Squalus mitsukurii Jordan and Snyder, 1903

DISTINGUISHING FEATURES: No anal fin. No lateral groove on dorsal fin spines. Often confused with *S. blainvillei* and *S. cubensis*. First dorsal fin over pectoral fins. Distance from fifth gill slit to first dorsal fin spine equal to distance from front of eye to second gill slit. Distance between dorsal fins greater than head length. Caudal peduncle with lateral keel. Teeth 17/23.

COLOR: Gray or brownish. No white spots on body or fins. Rear pectoral fins white edged.

DISTRIBUTION AND OCCURRENCE: North Carolina to Gulf of Mexico and off Argentina in 394-meter waters. North Carolina waters from shore to 366 meters deep; South Carolina from shore to 412 meters deep. Occurs in North Carolina June–July; reported in South Carolina (33°10′ N, 77°07′ W) in 308 meters as *S. fernandinus*.

SIZE: Attains lengths of 110 centimeters.

REFERENCES: Bass, D'Aubrey, and Kistanasamy 1976; Bearden 1965a, 1965b; Bigelow and Schroeder 1953, 1957; Compagno 1984; Garrick 1960b; McEachran and Fechhelm 1998; Merritt 1973; Munoz-Chapuli and Ramos 1989; Schwartz 1989b; Shirai 1992a; Wilson and Seki 1994.

Scalloped hammerhead

Sphyrna lewini (Griffith and Smith, 1834)

DISTINGUISHING FEATURES: Head flattened and wide, resembles hammer; center outer margin of broadly arched head notched with median indentation; outer lateral rear edge of mouth, viewed ventrally, ends behind corner of mouth. Distal edges of pectoral fins straight. First dorsal fin falcate. Origin of first dorsal fin over pectoral fins insertions. Second dorsal fin height less than that of anal fin. No dorsal ridge between fins. Teeth 15 or 16-0 to 2-15 or 16/15 or 16-1 to 2-15 or 16.

COLOR: Gray or with greenish-yellow sheen dorsally, white ventrally. Pectoral fin tips dusky.

DISTRIBUTION AND OCCURRENCE: New Jersey to Brazil in inshore waters to depths of 275 meters. Year-round resident in the Gulf Stream. Commonest hammerhead shark in Carolinas. Inshore continental shelf visitor March–September; even enters estuaries, rivers, and sounds. North Carolina river records are for Ward Creek, North River, and Cape Fear River, March–November. One North Carolina population migrates inshore-offshore to the Gulf Stream for the winter; another migrates seasonally north and south along the coast in shelf waters June–September.

SIZE: Attains 4.2-meter lengths. Largest North Carolina specimen was 3.1 meters captured 3 September 1976; South Carolina largest scalloped hammerhead (213 kilograms) was caught September 1976 off Edisto Island in 21-meter ocean depths.

REFERENCES: Bigelow and Schroeder 1948; Clark 1984; Compagno 1984; Garrick and Schultz 1963; C. R. Gilbert 1967a, 1967b; C. F. Jensen 1998;

Jensen and Hopkins 2001; Kohler, Casey, and Turner 1998; McKenzie 1970; Moore and Farmer 1981; Pratt et al. 1998; Schwartz 1993a, 2000a; Schwartz and Jensen 1995.

Great hammerhead

Sphyrna mokarran (Rüppell, 1837)

DISTINGUISHING FEATURES: Head wide, hammer shaped. Outer margin of head straight but with notch. Outer lateral rear edge of head, viewed ventrally, ends on level of mouth symphysis). High first dorsal fin strongly falcate, rear tip in front of pelvic fins origins. Rear edge of pectoral fins straight. Second dorsal fin height equals anal fin height. Teeth 17-2 or 3-17/16 or 17-1 to 3-16 or 17.

COLOR: Gray, lighter ventrally. May have bronze sheen in fall. Young with dark-tipped second dorsal fin.

DISTRIBUTION AND OCCURRENCE: Circumtropical. North Carolina to Brazil, including Gulf of Mexico and Caribbean to 80-meter depths. South Carolina visitor during summer months.

SIZE: Attains 6.1-meter size. North Carolina largest specimen 3.3 meters long, 162 kilograms captured 10 October 1975.

REFERENCES: Bigelow and Schroeder 1948; Compagno 1984; Garrick and Schultz 1963; C. R. Gilbert 1967a, 1967b; Kohler, Casey, and Turner 1998; S. Springer 1960, 1963.

Smooth hammerhead

Sphyrna zygaena (Linnaeus, 1758)

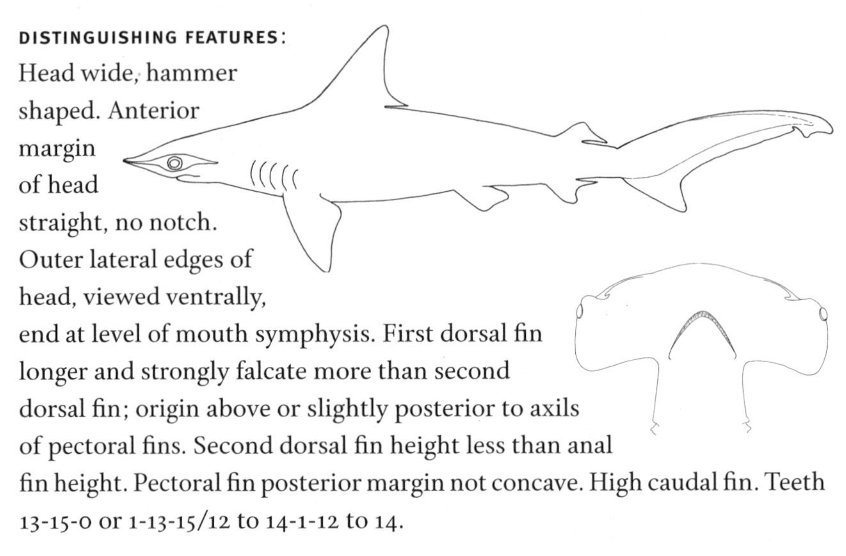

DISTINGUISHING FEATURES: Head wide, hammer shaped. Anterior margin of head straight, no notch. Outer lateral edges of head, viewed ventrally, end at level of mouth symphysis. First dorsal fin longer and strongly falcate more than second dorsal fin; origin above or slightly posterior to axils of pectoral fins. Second dorsal fin height less than anal fin height. Pectoral fin posterior margin not concave. High caudal fin. Teeth 13-15-0 or 1-13-15/12 to 14-1-12 to 14.

COLOR: Gray above to lighter ventrally. Young darker above.

DISTRIBUTION AND OCCURRENCE: Nova Scotia to northern Argentina, including Gulf of Mexico and Caribbean. Young may enter rivers; adults are usually offshore. Occur year-round in open warm ocean waters; occasionally in North Carolina June–October, in South Carolina May–September.

SIZE: Largest 2.3-meter fork length, 1 October 1975.

REFERENCES: Bigelow and Schroeder 1948; Compagno 1984; Garrick and Schultz 1963; C. R. Gilbert 1967a; Gudger 1907, 1932, 1947; C. F. Jensen 1998; Jordan and Gilbert 1882; Kohler, Casey, and Turner 1998; McEachran and Fechhelm 1998; Moore and Farmer 1981; Pratt et al. 1998; Radcliffe 1916; Smith 1907; S. Springer 1963; Wilson 1900.

Bonnethead shark

Sphyrna tiburo (Linnaeus, 1758)

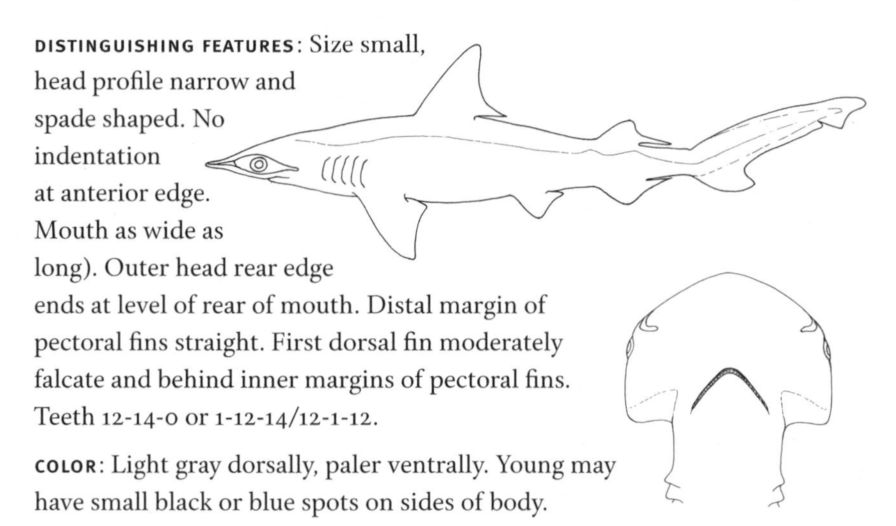

DISTINGUISHING FEATURES: Size small, head profile narrow and spade shaped. No indentation at anterior edge. Mouth as wide as long). Outer head rear edge ends at level of rear of mouth. Distal margin of pectoral fins straight. First dorsal fin moderately falcate and behind inner margins of pectoral fins. Teeth 12-14-0 or 1-12-14/12-1-12.

COLOR: Light gray dorsally, paler ventrally. Young may have small black or blue spots on sides of body.

DISTRIBUTION AND OCCURRENCE: Rhode Island, North Carolina to southern Brazil, Cuba, and Bahamas in waters to 80 meters deep. Enter sounds and inlets. Variably abundant in North Carolina from June to November in Core Sound, Bogue Sound, fresh waters of Neuse River, Cape Fear River; abundant South Carolina captures occur April to September to 80-meter depths.

SIZE: Attains 150-centimeter lengths in Carolinas and elsewhere.

REFERENCES: Bigelow and Schroeder 1948; Compagno 1984; C. R. Gilbert 1967a; Jordan and Gilbert 1882; Kohler, Casey, and Turner 1998; Myrberg and Gruber 1974; McEachran and Fechhelm 1998; Moore and Farmer 1981; Radcliffe 1916; Schwartz 2000a; Schwartz and Jensen 1995; Smith 1907.

Sand tiger shark

Carcharias taurus Rafinesque, 1810

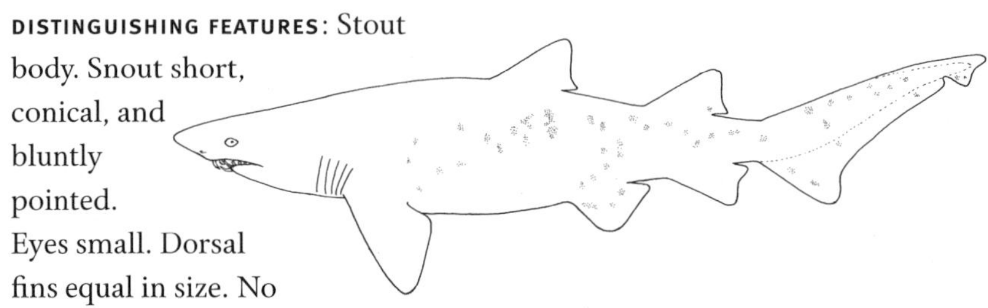

DISTINGUISHING FEATURES: Stout body. Snout short, conical, and bluntly pointed. Eyes small. Dorsal fins equal in size. No nictitating lower eyelids. Dorsal precaudal pit; pit absent ventrally. No interdorsal ridge. First dorsal fin behind midpoint of body. First dorsal fin closer to pelvic fins than to pectoral fins bases. Teeth protrude from jaws. No caudal keels. Teeth 36-54/32-46.

COLOR: Gray to lemon-yellow, white below. Gray spots scattered on body and tail.

DISTRIBUTION AND OCCURRENCE: Gulf of Maine to Florida, northern Gulf of Mexico to southern Brazil and Argentina in depths to 191 meters. Common year-round in Carolinas, especially July–November in shallow shelf waters. Usually found around wrecks from Cape Hatteras south.

SIZE: Attains lengths of 3.1 meters in Carolinas and elsewhere.

REFERENCES: Bigelow and Schroeder 1948, 1953; Compagno 1984; Jensen and Hopkins 2001; Jordan and Gilbert 1882; Kohler, Casey, and Turner 1998; Radcliffe 1916; Schwartz 1989b; S. Springer 1948; Smith 1907.

Smalltooth sand tiger

Odontaspis ferox (Risso, 1810)

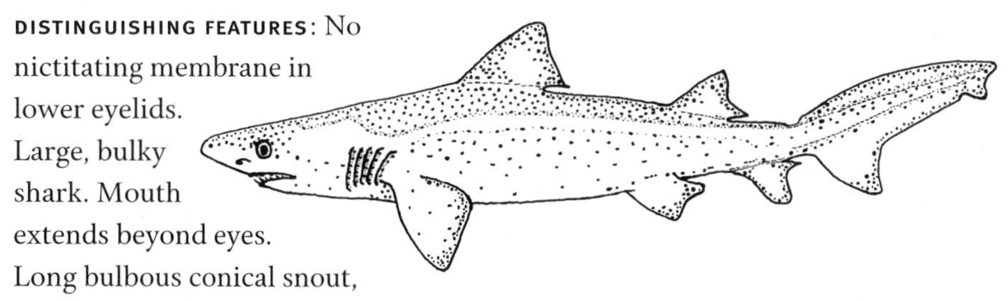

DISTINGUISHING FEATURES: No nictitating membrane in lower eyelids. Large, bulky shark. Mouth extends beyond eyes. Long bulbous conical snout, eyes large. Second dorsal and anal fins smaller than first dorsal fin. Second dorsal fin closer to pectoral fins than to pelvic fins. Upper precaudal pit present. No lateral keels on caudal peduncle. Upper lobe of caudal fin less than half of body length. Teeth with 2 or 3 lateral cusplets 23-23/17-17.

COLOR: Gray, lighter below. No spots on North Carolina specimen.

DISTRIBUTION AND OCCURRENCE: Western Atlantic and elsewhere deeper to 420-meter waters. A single 237-centimeter total length, 250-kilogram North Carolina specimen was captured 18 September 1994 off Cape Hatteras in 178-meter waters (34°51′N, 75°26′W). No South Carolina captures known.

SIZE: North Carolina specimen 2.4 meters total length.

REFERENCES: Compagno 1984, 2001; Schwartz 2000b; Sheehan 1998.

NO NICTITATING EYELID MEMBRANE, TAIL ONE-THIRD TO ONE-HALF OR MORE BODY LENGTH

Bigeye thresher

Alopias superciliosus (Lowe, 1839)

DISTINGUISHING FEATURES: Large oval eyes, higher than long, high on head. Head in adults grooved, giving helmet appearance. No nictitating lower eyelids. Pectoral fins large. No ridges or keels on body or tail. Precaudal pit present. Upper lobe of caudal fin nearly half body length. Teeth 11-11/10-10.

COLOR: Gray body to lighter below.

DISTRIBUTION AND OCCURRENCE: New Jersey to Cuba. Abundant off Cape Hatteras February–June. Earliest North Carolina capture 26–27 February 1963 off Cape Hatteras at 35°39′N, 74°27′W. One inshore record in Core Sound (Carteret County), May 1987, size 3.0 meters long.

SIZE: Specimens 3.2 meters long are sporadically common south of Shackleford Banks (Carteret County). A 2.3-meter long, 183-kilogram specimen was captured off Edisto Island, South Carolina, June 1978.

REFERENCES: Bigelow and Schroeder 1948; Compagno 1984, 2001; Gruber and Compagno 1983; Kohler, Casey, and Turner 1998; Moore and Farmer 1981; Schwartz 1989b; Stillwell and Casey 1976.

Thresher shark

Alopias vulpinus (Bonnaterre, 1788)

DISTINGUISHING FEATURES: Eyes nearly round, not ovate. No groove on head to give helmet appearance (see bigeye thresher). No nictitating eyelid membrane. Snout short, rounded. First dorsal fin close to pectoral fins. No ridges on body or tail. Upper lobe of caudal fin more than half body length. Pectoral fin falcate and pointed. Second dorsal fin origin before anal fin origin. Teeth 58+/58+.

COLOR: Gray dorsally, white ventrally with scattered intermixed black blotching over ventral surface.

DISTRIBUTION AND OCCURRENCE: Oceanic Newfoundland to Argentina. A February–August continental shelf inhabitant of the Carolinas. Year-round resident offshore off Cape Hatteras. Common in shallow waters along Bogue Banks in November as they migrate south.

SIZE: North Carolina specimens to 4.9 meters long. Rare in South Carolina.

REFERENCES: Bigelow and Schroeder 1948; Compagno 1984; Kohler, Casey, and Turner 1998; Moore and Farmer 1981; Schwartz 1989b.

WELL-DEVELOPED LATERAL KEELS ON CAUDAL PEDUNCLE

Basking shark

Cetorhinus maximus (Gunnerus, 1765)

DISTINGUISHING FEATURES: Size large. Enormous gill slits extend from top of head and nearly meet ventrally. Head conical, somewhat pointed. No dorsal fin spines. Pectoral fins short. First dorsal fin longer, more triangular than second. Eyes small on sides of head. Bristle-like gill rakers visible at gill slits. All gill slits are anterior to the pectoral fin base. Second dorsal and anal fins half size of first dorsal fin. Strong keels on caudal peduncle. Tail nearly lunate shaped. Teeth minute 35-63/34-68.

COLOR: Black body. Interior of mouth, which is held open as cruises at surface feeding on plankton, snowy white. Two albino specimens known from North Carolina.

DISTRIBUTION AND OCCURRENCE: Found from Gulf of Maine and Newfoundland, Gulf of Mexico, and in South America to Argentina. Frequents Carolina cold shelf waters and occasionally in Pamlico Sound November–May. Retreats northward as waters warm above 10°C (see Introduction). Enters Pamlico Sound, North Carolina, in May. Most common from Cape Hatteras to especially Cape Lookout area and Wrightsville Beach, North Carolina.

Rare in South Carolina. Over 360 occurrences known in North Carolina between 1901 and 2002 (Brimley 1935a; Gudger 1948a; Schwartz 2002).

SIZE: Two- to 6-meter lengths in North Carolina; larger (9.8 meters) in Arctic waters. Not affected by red tides.

REFERENCES: Bigelow and Schroeder 1948; Brimley 1935a; Castro 1996b; Compagno 1984, 2001; Gudger 1948a; Kohler, Casey, and Turner 1998; McEachran and Fechhelm 1998; Moore and Farmer 1981; Parker and Stott 1965; Schwartz 1978, 2002; Sims and Reid 2002; Springer and Gilbert 1976; Van der Molen, Caille, and Gonzalez 1998.

Great white shark

Carcharodon carcharias (Linnaeus, 1758)

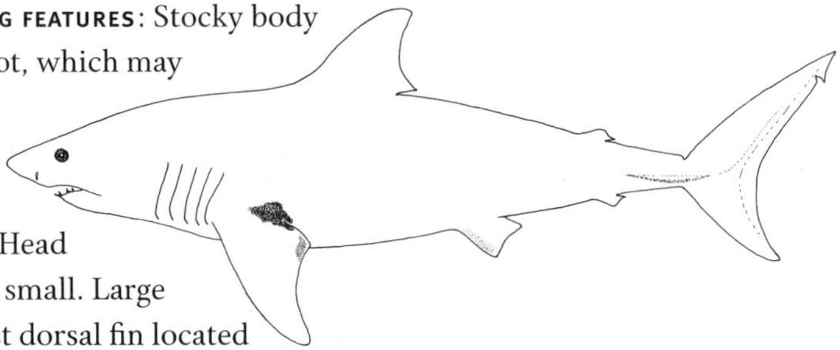

DISTINGUISHING FEATURES: Stocky body with black spot, which may fade in large individuals, in axils of pectoral fins. Head pointed. Eyes small. Large triangular first dorsal fin located above and slightly anterior to outer corner of pectoral fin. Pectoral fins broad and shorter than head. Strong single lateral caudal peduncle keel. Second dorsal fin in advance of anal fin. Caudal fin equally lobed and nearly lunate. Teeth 13-13/10-10.

COLOR: Gray dorsally, white ventrally. May have tinge of yellow ventrally.

DISTRIBUTION AND OCCURRENCE: Worldwide. Frequents Carolina inshore ocean coastal waters April and July–August as cold, deep waters intrude onto the continental shelf. Sporadic inshore September captures off Oregon Inlet, and in April, 3 kilometers south of Shackleford Banks. A 4.8-meter, 930-kilogram specimen captured at edge of shelf, 27 April 1986. Other 4.5-meter specimens have been captured in 36-meter waters off Hatteras. A 3.2-meter, 394-kilogram female was captured 6 kilometers southeast of Drum Inlet, 3 May 1996. A 4.5-meter specimen was seen 5 August 2000 just east of Cape

Lookout shoals in 23°C waters, depth 36 meters. Another similar-sized great white shark was seen at nearly the same location 1 July 2001. Rare in South Carolina. A dangerous shark that attacks humans.

SIZE: Maximum size 6.4 meters total length.

REFERENCES: Bigelow and Schroeder 1948, 1953; Casey and Pratt 1985; Compagno 1984, 2001; Coles 1919; Ellis and McCosker 1991; Garrick and Schultz 1963; Kohler, Casey, and Turner 1998; Moore and Farmer 1981; Mollet et al. 1996; Pratt et al. 1982; Randall 1973; Schwartz 1989b; Smale and Heemstra 1997.

Shortfin mako

Isurus oxyrinchus Rafinesque, 1810

DISTINGUISHING FEATURES: Snout long and pointed. Teeth often protrude. Pectoral fin short, narrow, pointed, and less than 70 percent of head length. Eyes large. First dorsal fin originates over rear tips of pectoral fins. Second dorsal fin small and located slightly ahead of anal fin. Broad keel on narrow caudal peduncle. No secondary keel on caudal fin base. Tail falcate and equally lobed. Teeth 12 or 13-12 or 13/12 or 13-12 or 13.

COLOR: Blue body. Snout and ventral head and belly white.

DISTRIBUTION AND OCCURRENCE: Worldwide. Georges Bank to Caribbean in waters to 152 meters deep. Common year-round in Gulf Stream. Supports active sport fisheries. Occurs in Carolina nearshore waters July–September; will jump when hooked. Considered a dangerous species in North Carolina.

SIZE: Attains total lengths of 3.9 meters.

REFERENCES: Benz and Caira 1995; Bigelow and Schroeder 1948; Compagno 1984, 2001; Garrick 1967b; Garrick and Schultz 1963; Kohler, Casey, and

Turner 1998; McEachran and Fechhelm 1998; Moore and Farmer 1981; Schwartz 1989b.

Longfin mako

Isurus paucus Guitart Manday, 1966

DISTINGUISHING FEATURES: Snout pointed. Mouth strongly arched. Small second dorsal fin located before anal fin. Pectoral fins narrow and broad tipped and as long as head. First dorsal fin large and originates over free rear tips of pectoral fins. Tail falcate and even-lobed. Broad keel on caudal peduncle. Secondary keel on caudal fin base. Teeth 24-26/24-26.

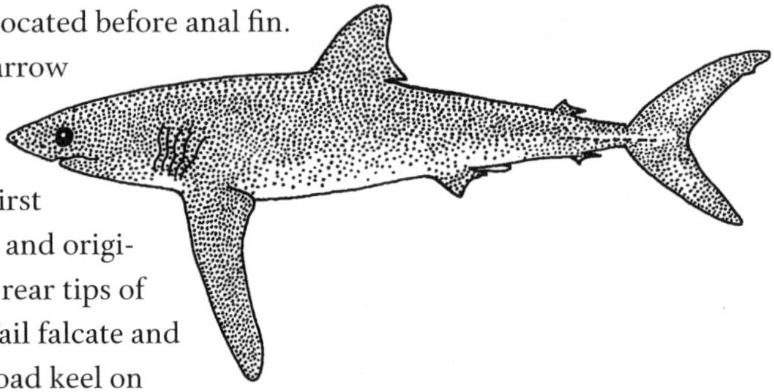

COLOR: Blue to black to dark gray ventrally. Snout and mouth area dusky.

DISTRIBUTION AND OCCURRENCE: Georges Bank. North Carolina off Cape Hatteras to Gulf of Mexico in depths of 220-meter water. North Carolina occurrences May–October. Off Cape Hatteras February–summer. South Carolina capture 27 May 1989 in 120-meter waters.

SIZE: Attains 4.2-meter lengths.

REFERENCES: Compagno 1984, 2001; Clark and Kristoff 1990; Dodril and Gilmore 1979; Guitart Manday 1966; Killam and Parsons 1986; Kohler, Casey, and Turner 1998; McEachran and Fechhelm 1998; Schwartz 1989b.

Porbeagle

Lamna nasus (Bonnaterre, 1788)

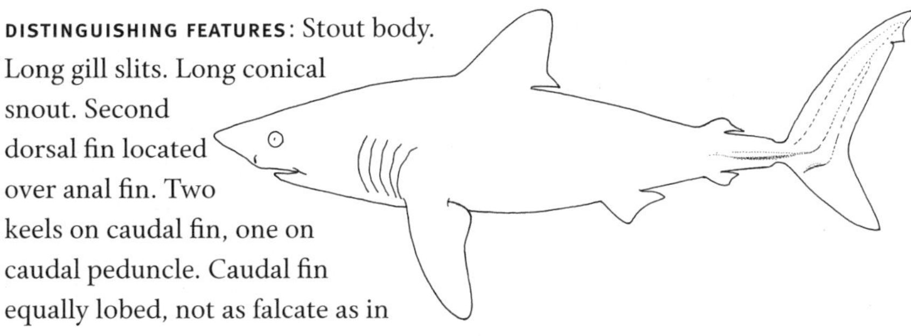

DISTINGUISHING FEATURES: Stout body. Long gill slits. Long conical snout. Second dorsal fin located over anal fin. Two keels on caudal fin, one on caudal peduncle. Caudal fin equally lobed, not as falcate as in mako or great white shark. Large first dorsal fin and small second dorsal and anal fins. Predorsal and ventral caudal pits. Teeth 12-15/12-14.

COLOR: Dark gray-white ventrally with dusky blotches. First dorsal fin rear base with white patch.

DISTRIBUTION AND OCCURRENCE: Newfoundland and Gulf of St. Lawrence to North Carolina; South Carolina rare. A commercial fishery in Canada and northern United States. Three specimens known from North Carolina off Cape Hatteras (J. Francisconi, personal communication, October 2001). Ocean occurrences to 36 meters deep.

SIZE: Attains 300-centimeter lengths.

REFERENCES: Bigelow and Schroeder 1948, 1953; Compagno 1984, 2001; Garrick and Schultz 1963; Jensen et al. 2002; Moore and Farmer 1981; Schwartz 1989b.

Whale shark

Rhincodon typus Smith, 1828

DISTINGUISHING FEATURES: Size large. Horizontal mouth at tip of snout. Eyes on sides of head. Spiracles smaller than eyes.

Head broad, flat. Body humped at pectoral fins. Three or 4 lateral ridges on body. Pectoral fins large. First dorsal fin larger than second. Precaudal pit present on upper caudal fin base. Distinct prominent keel on caudal peduncle. Nostrils with very small barbels. Long upper caudal fin lobe. Caudal fin less than one-third body length. Teeth more than 300 rows in either jaw.

COLOR: Gray-black; white spots over dorsal body, pectoral and caudal fins. Ten or more transverse stripes across body from level pectoral fins to caudal peduncle.

DISTRIBUTION AND OCCURRENCE: Cosmopolitan. Western Atlantic Ocean from New York to Brazil and Gulf of Mexico to 18-meter depths. Several (4) off Beaufort Inlet, North Carolina, in summer. Rare in South Carolina.
SIZE: Usually 12-meter lengths, perhaps longer. World's largest fish.

REFERENCES: Bigelow and Schroeder 1948; Brimley 1935b; Compagno 1984, 2001; Gudger 1915; Schwartz 1989b.

ANAL FIN PRESENT

Iceland cat shark

Apristurus laurussoni (Saemundsson, 1922)

DISTINGUISHING FEATURES: Snout short and wide. Gill filaments often exposed in gill slits. Small spiracles present. Flattened head. Slender; two dorsal fins, no spines in fins. Space between dorsal and caudal fins. Denticles on dorsal edge of caudal fin not conspicuous as a crest. Snout to pelvic fin distance greater than in *A. profundorum*. Interspace between pectoral and pelvic fin bases about 11 percent of total length in adults. Anal fin separate from caudal fin by notch. Teeth 42-0-41/53-0-43.

COLOR: Gray brown.

DISTRIBUTION AND OCCURRENCE: Massachusetts to Delaware and Gulf of Mexico to 1,462-meter depths. No records for Carolinas; expected in deep waters off edge of continental shelf.

SIZE: Attains 68-centimeter total length.

REFERENCES: Bigelow and Schroeder 1944; Bigelow, Schroeder, and Springer 1953; Compagno 1984; Nakaya 1991; Nakaya and Sato 1997; S. Springer 1966, 1979.

Deepwater cat shark

Apristurus profundorum (Goode and Bean, 1896)

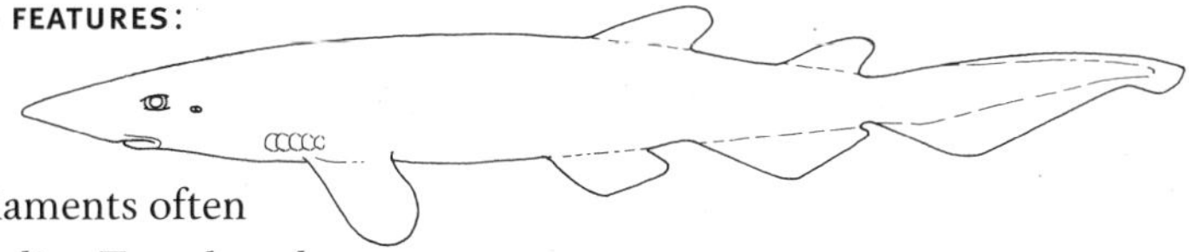

DISTINGUISHING FEATURES: Slender; snout long. Spiracles present. Gill filaments often exposed in gill slits. Two dorsal fins; no spines in fins. No space between second dorsal and caudal fin. Denticles on dorsal edge of caudal fin conspicuous as a crest. Distance from snout to origin of first dorsal fin less than in *A. laurussoni*. Pelvic fins short and bulbous. Interspace between pectoral and pelvic fin bases about 15 percent of total length in adults. No space between anal and caudal fins. Teeth 25-25/25-25.

COLOR: Brownish-black.

DISTRIBUTION AND OCCURRENCE: Delaware to Caribbean. Carolina record deep waters to 1,317 meters.

SIZE: Attains 51-centimeter lengths.

REFERENCES: Nakaya and Sato 1997; Nakaya and Stehmann 1998; S. Springer 1966, 1979.

Roughtail cat shark

Galeus arae (Nichols, 1927)

DISTINGUISHING FEATURES: Slender, long-pointed snout. Spiracles present. Space between second dorsal and caudal fins. Anal fin short. Distance from tip of snout to cloaca greater than remainder of body and tail. Caudal fin crest of denticles evident, dorsally, not present on lower side of caudal peduncle or leading edge of lower caudal fin lobe. Teeth 36-36/35-35.

COLOR: Brownish with dark spots and vertical saddlelike blotches on dorsal and lateral body surfaces. Dark streak from snout to eye. Pale ventrally. Mouth black.

DISTRIBUTION AND OCCURRENCE: North Carolina to Florida Keys, Gulf of Mexico and Belize to Cost Rica in 732-meter waters.

SIZE: Attains 433-centimeter lengths.

REFERENCES: Bigelow and Schroeder 1948; Bullis 1967; Compagno 1984; S. Springer 1966, 1979; Schwartz 1989b.

White-saddled cat shark

Scyliorhinus hesperius Springer, 1966

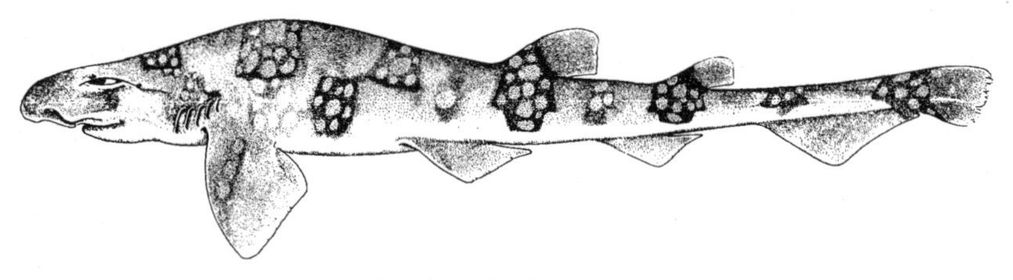

DISTINGUISHING FEATURES: Slender, dark-saddle-marked cat shark. Second dorsal fin much smaller than first. First dorsal fin behind pelvic fin insertion. Interdorsal fin space slightly larger or less than anal fin base. Resembles *S. torrei*, which ranges from south Florida to Cuba and Bahamas. Teeth 24-0-24/22-2-24.

COLOR: Tannish with 7 or 8 darker saddles with white spots interspersed in the darker saddles. Light areas between darker saddles with white spots. No black spots.

DISTRIBUTION AND OCCURRENCE: North Carolina and Caribbean to South America. Specimens captured off Carolinas in deep water depths to 457 meters.

SIZE: Attains 47-centimeter lengths.

REFERENCES: Compagno 1984; S. Springer 1966, 1979; Springer and Sadowsky 1970.

Blotched cat shark

Scyliorhinus meadi Springer, 1966

DISTINGUISHING FEATURES: Slender deep-water shark. Spiracles small. Second dorsal fin smaller than first. Distance from tip of snout to cloaca equal to remainder of body and tail. No enlarged denticles forming crest on upper lobe of caudal fin. Teeth 25-0-25/25-0-25.

COLOR: Gray with 7 distinct blotches or saddles, 2 of which extend onto both dorsal fins.

DISTRIBUTION AND OCCURRENCE: North Carolina southward into Caribbean. North Carolina specimens caught in July and October in 450- to 495-meter water depths. One specimen was 490 millimeters long caught off Cape Fear River. Another was 385-millimeter total length caught in 450-meter water 30 October 1966 (34°12.7′N, 75°11.5′W).

SIZE: Attains total lengths of 490 millimeters.

REFERENCES: Bigelow and Schroeder 1948; Burgess, Link, and Ross 1979; Compagno 1984; Parsons 1985; S. Springer 1966, 1979; Springer and Sadowsky 1970.

Chain cat shark

Scyliorhinus retifer (Garman, 1881)

DISTINGUISHING FEATURES: Slender. Spiracles present. Space exists between second dorsal and caudal fins. First dorsal fin origin slightly posterior to pectoral fins bases. Distance from tip of snout to cloaca equals remainder of body and tail length. No enlarged denticles forming crest on upper lobe of caudal fin. Teeth 21-0-21/20-1-20.

COLOR: Tan-brown, often with yellow sheen. Brown/black chainlike markings or bars on body and dorsal fins. No light or dark spots on bars. Seven or 8 saddle markings. Teeth 46-52/46.

DISTRIBUTION AND OCCURRENCE: Cape Cod to Florida and Caribbean. North Carolina specimens captured July and October in 42- to 53- and 450- to 495-meter waters southeast of Cape Lookout year-round. Also off Nags Head, North Carolina, in deep water off continental shelf. Year-round resident in very deep waters.

SIZE: Attains total lengths of 47 centimeters.

REFERENCES: Able and Flescher 1991; Bigelow and Schroeder 1948; Bigelow, Schroeder, and Springer 1953; Compagno 1984; McEachran and Fechhelm 1998; Radcliffe 1916: Schwartz 1989b; S. Springer 1966, 1979; Springer and Sadowsky 1970; Smith 1907.

PRECAUDAL PITS ABSENT, NICTITATING EYELIDS PRESENT, INTERDORSAL RIDGE PRESENT

Dusky smooth hound (smooth dogfish)

Mustelus canis (Mitchell, 1815)

DISTINGUISHING FEATURES: Slender, body somewhat triangular shaped in cross section. Short head; eyes large, slitlike-oval. Mouth slightly longer than eye length. Both dorsal fins nearly equal in size. Second dorsal fin ahead of anal fin. No spines in dorsal fins. Ridge between dorsal fins. Anal-caudal fin space greater or equal in height to second dorsal fin. Lower lobe of caudal fin does not project rearward. Teeth 74/80.

COLOR: Uniform gray above, lighter below. No white spots or bars on body.

DISTRIBUTION AND OCCURRENCE: New Brunswick, Canada, to southern Brazil. Off Carolinas year-round usually in less than 274-meter ocean waters. Occurs in large schools, October–June. Supports commercial fishery along entire Carolina coasts; however, greater numbers occur from Cape Lookout northward. Enters estuaries, sounds, and rivers. Pups in late April. Adults may remain in coastal waters until July, when waters warm, causing northward migrations. Contribute to pier fisheries May–July. Pups very common in Bull's Bay, South Carolina, and along North Carolina coast.

SIZE: Attains 150-centimeter lengths; females larger than males.

REFERENCES: Bigelow and Schroeder 1940, 1948; Castro 1993b; Compagno 1984; Coles 1926; Heemstra 1997; C. F. Jensen 1998; Jensen and Hopkins 2001; Jordan and Gilbert 1882; Moore and Farmer 1981; Schwartz 1995, 2000a; Schwartz and Porter 1977; Schwartz, Hogarth, and Weinstein 1982; Smith 1907; S. Springer 1939.

Florida smooth hound

Mustelus norrisi Springer, 1940

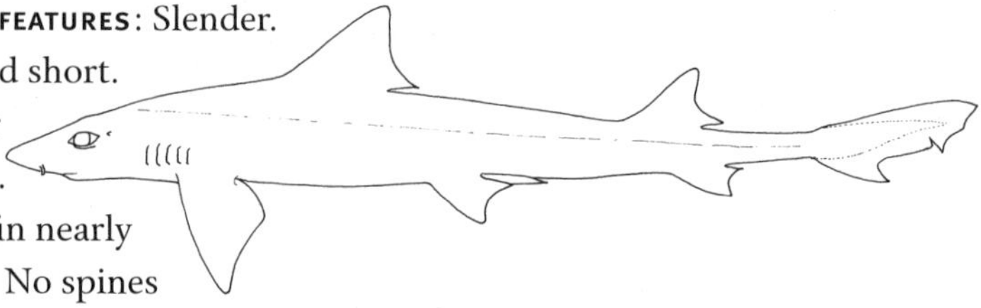

DISTINGUISHING FEATURES: Slender. Large eyes. Head short. Body triangular in cross section. Second dorsal fin nearly as large as first. No spines in dorsal fins. Ridge between dorsal fins. Ventral lobe of caudal fin pointed rearward. First dorsal fin located posterior to inner margin of pelvic fin. Interdorsal space about one-fourth of total length. Teeth 58-65/57-60.

COLOR: Gray above, no white spots or dark lines.

DISTRIBUTION AND OCCURRENCE: South Carolina southward to Florida, Gulf of Mexico, Mexico to southern Brazil. Found in 80-meter waters in South Carolina. Rare in North Carolina.

SIZE: Attains 100-centimeter lengths.

REFERENCES: Bigelow and Schroeder 1948; Compagno 1984; Heemstra 1997; S. Springer 1939.

RIDGE PRESENT BETWEEN DORSAL FINS

Bignose shark

Carcharhinus altimus (Springer, 1950)

DISTINGUISHING FEATURES: Slender. Snout big, blunt; length equal to or greater than mouth width. Mouth extends behind level of eye. First dorsal fin behind axils of pectoral fins, over in *C. plumbeus*. Interdorsal ridge between dorsal fins. High caudal fin lobe larger than lower lobe. Second dorsal fin less than half first dorsal fin and origin before anal fin. Internal nictitating eyelids. Precaudal pits present. Teeth 14-16/14-15.

COLOR: Gray to bronze. White ventrally. Fin tips may be dusky.

DISTRIBUTION AND OCCURRENCE: North Carolina to Venezuela in 430-meter waters. North Carolina inshore along coast and offshore off Cape Hatteras to 225-meter water depths. Often migrate vertically in water column.

SIZE: Attains 282-centimeter lengths.

REFERENCES: Anderson and Stevens 1996; Bass, D'Aubrey, and Kistanasamy 1973; Bigelow and Schroeder 1948; Compagno 1984; Garrick 1982; Kohler, Casey, and Turner 1998; Schwartz 1989b; S. Springer 1950b.

Night shark

Carcharhinus signatus (Poey, 1868)

DISTINGUISHING FEATURES: Head long, pointed, one-third of body length. Eyes bright green in life, larger and not green in *C. altimus*. Nictitating eye membranes, no spiracle. Snout narrow, equal to or greater than mouth width. Short gill slits; longest 2.5 percent of total body length. First dorsal fin above free tips of pectoral fins. Second dorsal fin less than height of first. Ridge between dorsal fins. Precaudal pits. Earlier known as *Hypoprion bigelowi*. Teeth 15-15/15-15.

COLOR: Gray-blue above to gray-white ventrally.

DISTRIBUTION AND OCCURRENCE: Delaware south to Brazil in deep water to 600 meters. North Carolina specimens usually frequent waters 221–540 meters deep from Cape Lookout to Southport (Brunswick County). March–September specimens captured in 20-meter waters. Rare in South Carolina.

SIZE: Attains 280-centimeter lengths.

REFERENCES: Bigelow and Schroeder 1948; Compagno 1984; Kohler, Casey, and Turner 1998; McEachran and Fechhelm 1998; Moore and Farmer 1981; Raschi, Munro, and Compagno 1982.

Silky shark

Carcharhinus falciformis (Bibron, 1839)

DISTINGUISHING FEATURES: Slender. Snout length about equal to mouth width. First dorsal fin leading edge curved, tip rounded. Second dorsal fin over anal fin. Free tip of both dorsal fins long. Ridge between dorsal fin bases. No keels. Predorsal caudal pit. Caudal peduncle slender, rounded. Teeth 14-16/13-7.

COLOR: Gray above to white below. Tip fins dusky. Once known as *Carcharhinus floridanus*. Light color on sides between pectoral and pelvic fins. No spots on body.

DISTRIBUTION AND OCCURRENCE: New England to Brazil. Year-round resident offshore in waters to 500 meters deep. Frequents inshore Carolina waters May–November; even enters sounds and Cape Fear River waters of 28 parts per thousand. Supports commercial fishery. Fisheries affected by federal season, sizes, catch quotas, and regulations.

SIZE: Attains 2.5-meter lengths.

REFERENCES: Bass, D'Aubrey, and Kistanasamy 1973; Bigelow and Schroeder 1948; Compagno 1984; Garrick 1967a, 1982; Garrick and Schultz 1963; Garrick, Backus, and Gibbs 1964; Schwartz 1984, 2000a; S. Springer 1950b.

Galapagos shark

Carcharhinus galapagensis (Snodgrass and Heller, 1905)

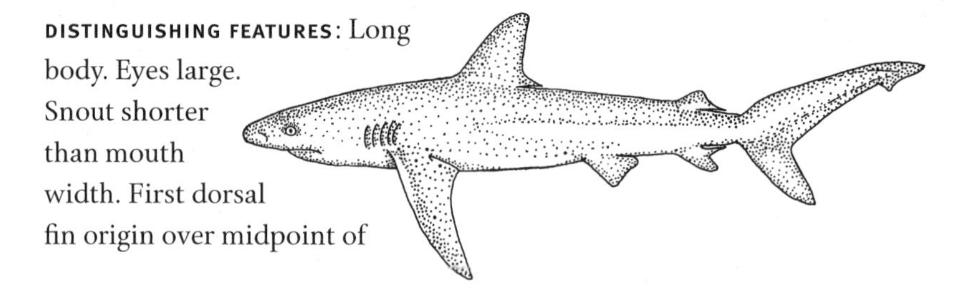

DISTINGUISHING FEATURES: Long body. Eyes large. Snout shorter than mouth width. First dorsal fin origin over midpoint of inner pectoral fin margins. Interdorsal ridge low. High large dorsal fin. Light "Z" coloration on sides of body between pectoral and pelvic fins. Second dorsal fin slightly ahead of anal fin. Closely resembles dusky shark but has taller second dorsal fin (2.6 to 2.8 percent of total length versus 1.5 to 2.3 percent in *C. obscurus*). Teeth 13-15-13-15/13-15-13-15.

COLOR: Dark gray above, white below. Fin tips dusky.

DISTRIBUTION AND OCCURRENCE: Worldwide, usually around oceanic islands. A single female 1.5 meters total length was captured off Southport (Brunswick County), North Carolina, 7 April 1990. No South Carolina records.

SIZE: Attains 3.7-meter lengths.

REFERENCES: Bass, D'Aubrey, and Kistanasamy 1973; Compagno 1984; Garrick 1967a, 1982; Garrick and Schultz 1963; Kohler, Casey, and Turner 1998; Schwartz 1998, 2000b.

Oceanic whitetip shark

Carcharhinus longimanus (Poey, 1861)

DISTINGUISHING FEATURES: Mouth twice as wide as long. Eyes small. Broad rounded first dorsal fin. First dorsal fin positioned over rear pectoral fins. Second dorsal fin small, above anal fin. Pectoral fin narrow, twice as broad, convex. Caudal peduncle round. Pectoral fin longer than distance from tip of snout to last gill slit. Ridge between dorsal fins. Precaudal pits. No caudal peduncle keels. Some adults may be ridgeless between dorsal fins. Teeth 14-2-14/14-1-14.

COLOR: Light gray to yellowish sheen to white below. Tips of dorsal, pectoral, pelvic, and caudal fins white. Fins may have black tips in young. A dark saddle on caudal peduncle.

DISTRIBUTION AND OCCURRENCE: Georges Banks to Uruguay. Year-round resident in open ocean and Gulf Stream and found in ocean waters 183 meters deep. A very aggressive and inquisitive shark.

SIZE: Attains 4.0-meter lengths.

REFERENCES: Backus, Springer, and Arnold 1956; Bass, D'Aubrey, and Kistanasamy 1973; Bigelow and Schroeder 1948; Compagno 1984; Garrick 1967a, 1982; Garrick and Schultz 1963; Kohler, Casey, and Turner 1998; Moore and Farmer 1981; Schwartz 1989b; S. Springer 1950b.

Dusky shark

Carcharhinus obscurus (Lesueur, 1818)

DISTINGUISHING FEATURES: Eyes large. Slender, similar to silky shark, but first dorsal fin is triangular shaped. First dorsal fin semifalcate, located over or slightly anterior to free rear tip of pectoral fins. Second dorsal fin origin over anal fin origin. Pectoral fins large and falcate; tips may be dusky but not black. Ridge between dorsal fins. No keels on caudal peduncle. White "Z" coloration on sides between pectoral and pelvic fins. Teeth 14-1 or 2-14/14-1-14.

COLOR: Gray, white ventrally, with bronze body sheen, often in fall.

DISTRIBUTION AND OCCURRENCE: Massachusetts to southern Brazil and Gulf of Mexico. Once most abundant in the Carolinas but numbers have decreased slightly in recent years. Common south of Cape Hatteras year-round. Penetrates estuaries and rivers. Common June–October in North Carolina, May–September in South Carolina. Adults may enter Bogue Sound, where a 909-millimeter-total-length shark was captured 1 August 1968. A 1.2-meter-total-length female with a deformed head was captured off Shackleford Banks, North Carolina, 24 October 1975. Frequents offshore water to 400-meter depths.

SIZE: Attains 4.0-meter lengths.

REFERENCES: Bass, D'Aubrey, and Kistanasamy, 1973; Bigelow and Schroeder 1948; Compagno 1984; Garrick 1982; Garrick and Schultz 1963; Huish and Benedict 1977; C. F. Jensen 1998; Jensen and Hopkins 2001; Moore and Farmer 1981; Pratt et al. 1998; Schwartz 1995, 2000a, 2000b; Schwartz, Hogarth, and Weinstein 1982; Smith 1907; S. Springer 1960, 1963.

Sandbar shark

Carcharhinus plumbeus (Nardo, 1827)

DISTINGUISHING FEATURES: Snout length less than mouth width. Gill openings short. Long narrow pectoral fin twice as long as height of first dorsal fin. First dorsal fin high, triangular, apex rounded, and situated over axil of pelvic fin. Second dorsal fin over anal fin. Interdorsal fin ridge present. Lower caudal fin lobe half upper lobe. Caudal peduncle round. No caudal peduncle keels. Once known as *Carcharhinus milberti*. Pups in area in May–June. Teeth 14-1-14/13 or 14-1-13 or 14.

COLOR: Blue-gray above (seasonally bronze in fall). White "Z" coloration on side between pectoral and pelvic fins.

DISTRIBUTION AND OCCURRENCE: New England to southern Brazil. Common in North Carolina south of Cape Hatteras June–September. Common shark in area. A 3.3-meter adult was captured in Newport River (Carteret County), 21 October 1907; otherwise an inshore ocean species. Young are common pier fishery captures in May–July and are called "sand sharks." Will enter fresh waters of sounds and rivers. Offshore found in waters to 30 meters deep.

SIZE: Attains 3.3-meter lengths.

REFERENCES: Bigelow and Schroeder 1948; Compagno 1984; Garrick 1967a, 1982; Garrick and Schultz 1963; C. F. Jensen 1998; Jensen and Hopkins 2001; Moore and Farmer 1981; Pratt et al. 1998; Schwartz 1989b, 2000a, 2000b; Sminkey and Musick 1996; Smith 1907; S. Springer 1960, 1963.

Caribbean reef shark

Carcharhinus perezi (Poey, 1876)

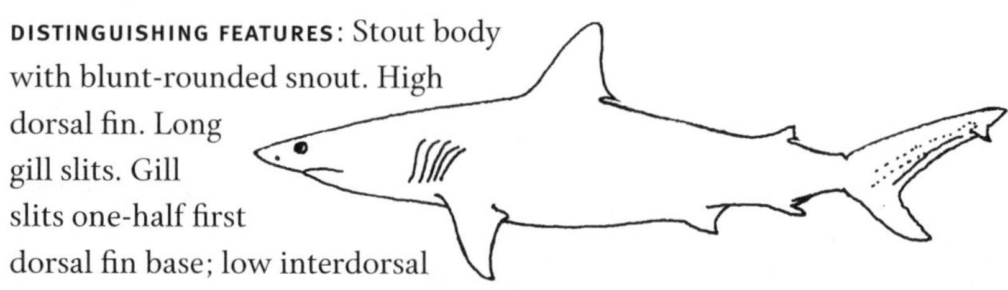

DISTINGUISHING FEATURES: Stout body with blunt-rounded snout. High dorsal fin. Long gill slits. Gill slits one-half first dorsal fin base; low interdorsal ridge present. First dorsal fin over or slightly anterior to pectoral free rear tips. No caudal peduncle keels. Often confused with bignose, dusky, Galapagos, and sandbar sharks. Once placed in genus *Eulamia* or known as *C. springeri*. Teeth 12 or 13-1 or 2-12 or 13/11 or 12-11 or 12.

COLOR: Gray above to white below. No "Z" coloration on sides of body between pectoral and pelvic fins. Undersides of paired fins dusky, not black.

DISTRIBUTION AND OCCURRENCE: North Carolina south to Brazil. A 1.9-meter-total-length, 45-kilogram male was captured 20 July 1991 at 35°13.4′N, 75°28′W in water 22 meters deep, 1 kilometer east of Cape Hatteras lighthouse.

SIZE: Attains 2.9-meter lengths.

REFERENCES: Bigelow and Schroeder 1948; Bonfil 1989; Compagno 1984; Garrick 1967a, 1982; Schwartz 2000b; Schwartz, Jensen, and Hopkins 1995; S. Springer 1960.

Tiger shark

Galeocerdo cuvier (Peron and Lesueur, 1822)

DISTINGUISHING FEATURES: Stout body. Gill slits short. Snout short and broadly rounded. Head flattened above. Small spiracle. High dorsal fin. Low prominent ridge between dorsal fins. Low keels on caudal peduncle. Upper lobe of caudal fin long, pointed, one-third of body length. Teeth 9-11-0-2-9-12/8-11-2-8-11.

COLOR: Blue-gray above, white below. Young lighter colored with prominent spots and bars on body and caudal fin. Spots and bars fade as maximum size attained.

DISTRIBUTION AND OCCURRENCE: Worldwide in tropical and temperate seas. Massachusetts to Uruguay, Gulf of Mexico, and Caribbean. Inshore to open ocean in Carolinas. Inshore May–November in North Carolina, April–September in South Carolina. Enter large rivers in North Carolina. A very dangerous shark, feeds on anything.

SIZE: Attains 7.4-meter lengths. Largest North Carolina specimen 3.8 meters captured 17 May 1987 and to 450 kilograms, especially off Southport (Brunswick County).

REFERENCES: Bigelow and Schroeder 1948; Clark and Kristoff 1990; Coles 1919; Compagno 1984; Garrick and Schultz 1963; Gudger 1932, 1948b, 1949; Kohler, Casey, and Turner 1998; McEachran and Fechhelm 1998; Moore and Farmer 1981; Radcliffe 1916; Randall 1973; Rider, Athorn, and Bailey 2002; Schwartz 1996b, 1998; S. Springer 1960, 1963.

NO RIDGE BETWEEN DORSAL FINS, DORSAL FIN AT OR BEFORE MIDPOINT OF BODY

Blacknose shark

Carcharhinus acronotus (Poey, 1860)

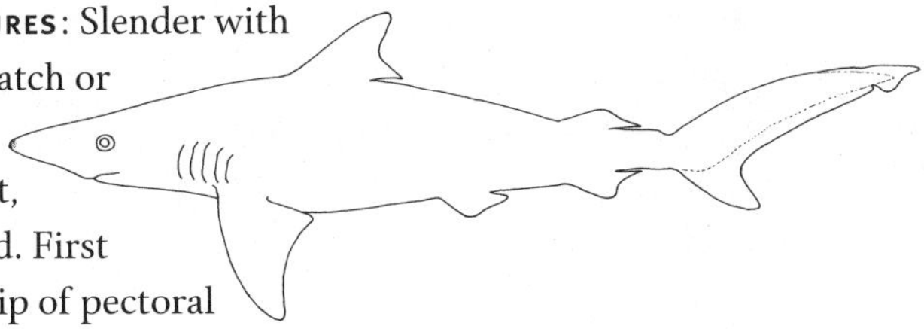

DISTINGUISHING FEATURES: Slender with conspicuous black patch or smudge at tip of snout. Gill slits short, less than half of head. First dorsal fin over free tip of pectoral fins. Second dorsal fin over anal fin. No caudal keels. No interdorsal ridge between fins. Teeth 12-13/11-12.

COLOR: Gray with yellow overcast or sheen (hence often identified as lemon shark); bronze body overcast in fall, white below from October into winter.

DISTRIBUTION AND OCCURRENCE: North Carolina to Brazil. Most common smaller shark in Carolinas May–October. Transient to Cape Hatteras. North Carolina specimens inshore to 280-meter waters. Will enter estuaries.

SIZE: Usually attains 1.1-meter lengths. North Carolina specimens of 1,830- and 2,055-millimeter total lengths captured 27 October 1975.

REFERENCES: Bigelow and Schroeder 1948; Compagno 1984; Garrick 1982; Gudger 1913a; Kohler, Casey, and Turner 1998; Radcliffe 1913, 1916; Schwartz 1989a, 1995, 2000a.

Finetooth shark

Carcharhinus isodon (Valenciennes, 1839)

DISTINGUISHING FEATURES: Eyes large. Slender shark with long, pointed snout. May be confused with *C. acronotus* but lacks a dark patch on tip of snout. Gill opening very long, one-half dorsal fin base. Second dorsal fin less than half of first and originates above anal fin. No ridge between dorsal fins. No caudal peduncle keels. Caudal peduncle round. Teeth 12-15/13-14.

COLOR: Metallic gray above, white below. Prominent light "Z" coloration on side of body between pectoral and pelvic fins.

DISTRIBUTION AND OCCURRENCE: New York to Gulf of Mexico to southern Brazil. Variably common in inshore waters in Carolinas, April–October in North Carolina, May–September in South Carolina. Abundance varies; some years very common in late summer, others rare: abundant in 1999 in North Carolina, 2000 in South Carolina.

Young were seen in inlets from Ocracoke to Oak Island, North Carolina, in 1998; otherwise only adults were captured. Bull's Bay seems to serve as a pupping nursery area in South Carolina, near Cape Hatteras in North Carolina.

SIZE: Attains 1.4-meter lengths.

REFERENCES: Bigelow and Schroeder 1948; Castro 1993a, 1993b; Compagno 1984; C. F. Jensen 1998; Pratt et al. 1998; Radcliffe 1916.

Spinner shark

Carcharhinus brevipinna (Müller and Henle, 1839)

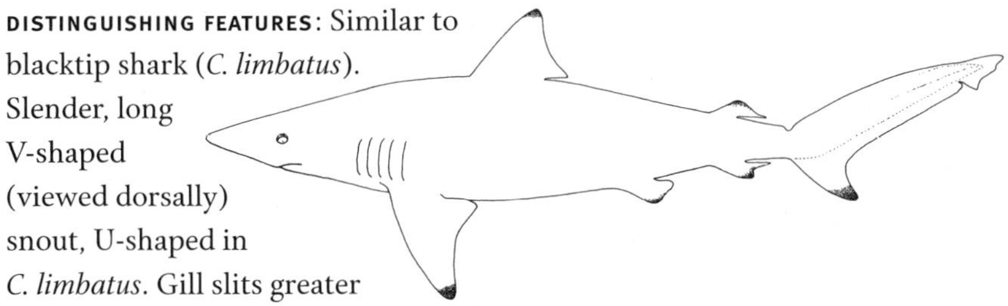

DISTINGUISHING FEATURES: Similar to blacktip shark (*C. limbatus*). Slender, long V-shaped (viewed dorsally) snout, U-shaped in *C. limbatus*. Gill slits greater than one-third first dorsal fin base. No interdorsal ridge. No caudal peduncle keel. Pectoral fins falcate and pointed and tips rounded. First dorsal fin over or behind inner corner of pectoral fin. Second dorsal fin less than half first dorsal fin height. Teeth 15-18/14-17.

COLOR: Gray, often with bronze sheen in fall. White below. Tips of dorsal, pectoral, anal, and lower caudal fins black. Whitish "Z" coloration on sides of body.

DISTRIBUTION AND OCCURRENCE: A common July–October inshore inhabitant of Carolina coastal waters. Most common from Cape Lookout, North Carolina, southward in water to 78 meters deep. Moves southward in September or when waters cool; will enter rivers. Makes spectacular jumps.

SIZE: Attains 1.5-meters lengths.

REFERENCES: Bass, D'Aubrey, and Kistanasamy 1973; Bigelow and Schroeder 1948; Branstetter 1982; Compagno 1984; Garrick 1982; Garrick and Schultz 1963; Kohler, Casey, and Turner 1998; Schwartz 1989b, 2000a; S. Springer 1960, 1963.

Bull shark

Carcharhinus leucas (Valenciennes, 1839)

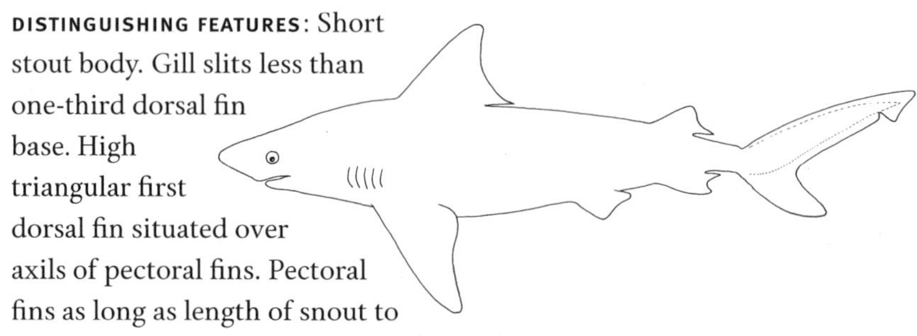

DISTINGUISHING FEATURES: Short stout body. Gill slits less than one-third dorsal fin base. High triangular first dorsal fin situated over axils of pectoral fins. Pectoral fins as long as length of snout to origin of pectoral fins. Broad U-shaped snout. Lower caudal fin lobe less than half upper lobe. Caudal peduncle rounded. No interdorsal fin ridge. Pups in North Carolina in May or June. Teeth 13-1-13/12-1-12.

COLOR: Dirty gray to lighter gray ventrally. Young have black caudal fin edge.

DISTRIBUTION AND OCCURRENCE: Worldwide. Often common in Carolinas, especially June–October in rivers, inlets, and sounds. Pups in North Carolina in May or June. Moves north and south seasonally as waters warm and cool. Most dangerous shark, causing several "shark attacks."

SIZE: Attains 3.4-meter lengths; in North Carolina 2.4 meters long.

REFERENCES: Bass, D'Aubrey, and Kistanasamy 1973; Bigelow and Schroeder 1948; Compagno 1984; Garrick 1982; Garrick and Schultz 1963; Jensen and Hopkins 2001; Kohler, Casey, and Turner 1998; Moore and Farmer 1981; Schwartz 1973, 1989b, 1995, 2000a; S. Springer 1960, 1963.

Blacktip shark

Carcharhinus limbatus (Valenciennes, 1839)

DISTINGUISHING FEATURES: Eyes more than one-fifth gill slit length. Often confused with spinner shark. Head short, U-shaped versus V-shaped in spinner shark. No interdorsal ridge. No caudal peduncle keels. Teeth 14-15-1-3-14-15/13-15-1-2-13-15.

COLOR: Gray above to white below. Whitish "Z" coloration on sides of body between pectoral and pelvic fins. All fins but anal fin black or dusky.

DISTRIBUTION AND OCCURRENCE: Massachusetts to Brazil. An inshore shelf-water visitor May–September in Carolinas in waters to 30 meters deep. Pups in July–August. Enters Core Sound and Cape Fear River in North Carolina. Spectacular jumper. Responsible for several "shark attacks."

SIZE: Attains lengths of 2.3 meters.

REFERENCES: Bass, D'Aubrey, and Kistanasamy 1973; Bigelow and Schroeder 1984; Branstetter 1982; Castro 1996a, 1996b; Compagno 1984; Garrick 1967a, 1982; Garrick and Schultz 1963; Gudger 1913a, 1932; C. F. Jensen 1998; Jensen and Hopkins 2001; Kohler, Casey, and Turner 1998; Pratt et al. 1998; Radcliffe 1916; Schwartz 1973, 1993a, 2000a; Schwartz, Hogarth, and Weinstein 1982; S. Springer 1960, 1963.

Lemon shark

Negaprion brevirostris (Poey, 1868)

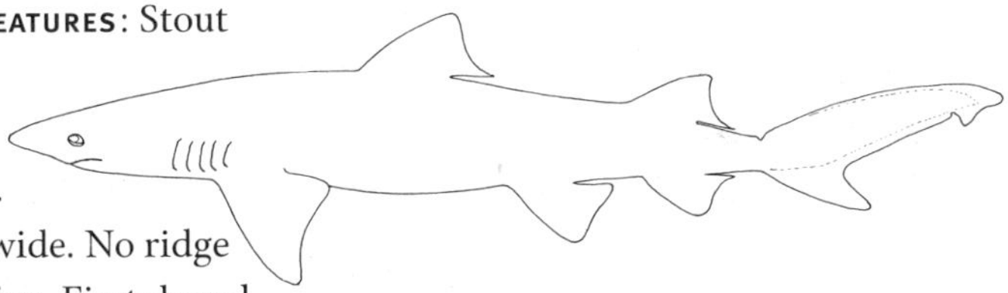

DISTINGUISHING FEATURES: Stout shark. Two large dorsal fins of nearly equal size. Head blunt and wide. No ridge between dorsal fins. First dorsal fin above or slightly posterior to free tip of pectoral fins. Second dorsal fin four-fifths height of first. No caudal peduncle keels. Precaudal dorsal pit present, absent ventrally. Teeth 15-1 or 3-15/13 or 14-3-13 or 14.

COLOR: Lemon-yellow to brownish body, lighter ventrally.

DISTRIBUTION AND OCCURRENCE: New Jersey to Brazil. Young and adults often enter sounds (Pamlico, Bogue, Core), estuaries (Broad Creek, Ward Creek, Calico Bay, Long Bay), and rivers (New River, Cape Fear River) in North Carolina in July–October. Rare in South Carolina. Frequents offshore waters to 92 meters deep.

SIZE: Attains lengths of 340 centimeters.

REFERENCES: Bigelow and Schroeder 1948; Compagno 1984; Moore and Farmer 1981; Moss 1967; Schwartz 2000a; Schwartz, Hogarth, and Weinstein 1982; S. Springer 1950a.

Blue shark

Prionace glauca (Linnaeus, 1758)

DISTINGUISHING FEATURES: Slender shark. Eyes large. First dorsal fin well behind pectoral fins. Long, rounded snout, long falcate pectoral fins.

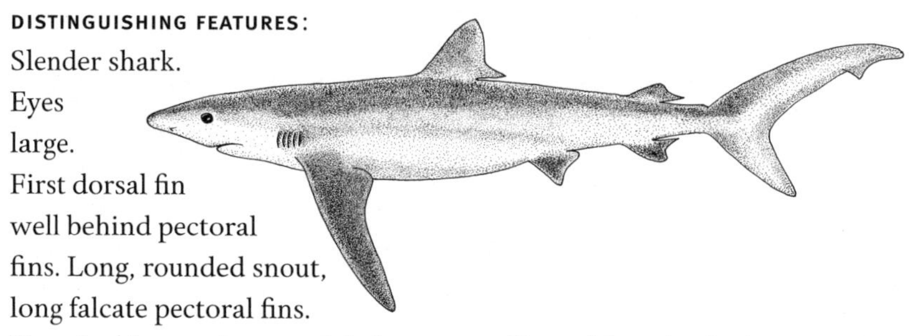

Dorsal and ventral precaudal pits present. No middorsal ridge between dorsal fins. Weakly developed lateral ridge on caudal peduncle. Upper caudal fin lobe twice lower lobe length. Slender caudal peduncle. Teeth 14-1-14/13-15-1-15-13.

COLOR: Blue-gray to black above, white below.

DISTRIBUTION AND OCCURRENCE: Newfoundland to Brazil in open ocean to depths over 300 meters. Year-round near surface off Carolinas.

SIZE: Attains 2.8-meter lengths. Largest North Carolina specimen captured 27 February 1982, 2.9 meters.

REFERENCES: Able and Flescher 1991; Bigelow and Schroeder 1948; Compagno 1988; Litvinov 1982, 1983; McEachran and Fechhelm 1998; Moore and Farmer 1981; Springer and Sadowsky 1970.

Atlantic sharpnose shark

Rhizoprionodon terraenovae (Richardson, 1836)

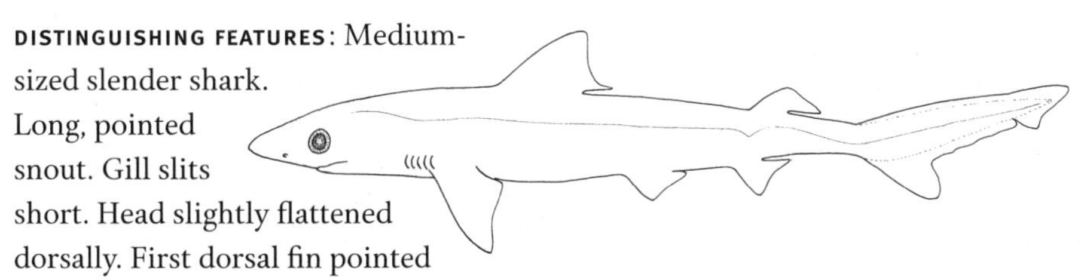

DISTINGUISHING FEATURES: Medium-sized slender shark. Long, pointed snout. Gill slits short. Head slightly flattened dorsally. First dorsal fin pointed and located above rear tip of pectoral fins. Second dorsal fin is located at midpoint of anal fin. No interdorsal ridge between fins. No caudal peduncle keels. Teeth 12-1-12/12-12.

COLOR: Gray above, white below. White spots scattered over body; white edges to pectoral fins, dorsal fins dusky tips.

DISTRIBUTION AND OCCURRENCE: Canada to Brazil and Gulf of Mexico from shore to 280-meter waters. Most common and abundant small shark in the Carolinas. Enters sounds, rivers, inshore to offshore. Pups in North Carolina in May–June; Bull's Bay in South Carolina serves as nursery area, as does area near Cape Hatteras, North Carolina.

SIZE: Attains 110-centimeter lengths.

REFERENCES: Bigelow and Schroeder 1948; Castro 1993b; Compagno 1984; Jenkins 1887; Jensen and Hopkins 2001; Jordan and Gilbert 1882; Kohler, Casey, and Turner 1998; Pratt et al. 1998; Radcliffe 1916; Schwartz 1984, 1993a, 1995, 2000a; V. G. Springer 1964.

skates and rays

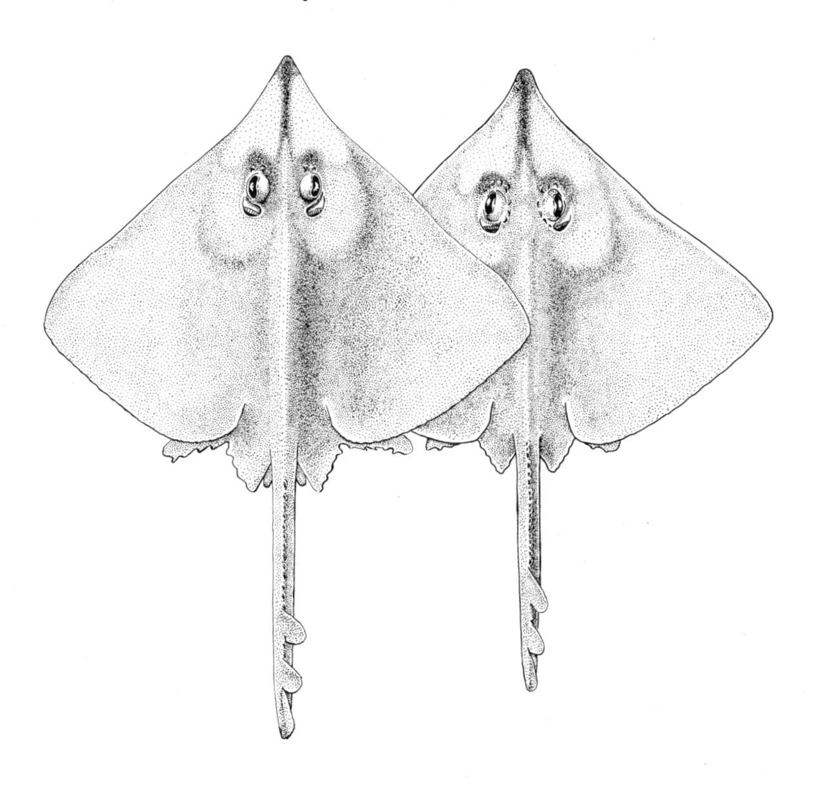

SNOUT BLADELIKE, ARMED WITH LATERAL TEETH

Smalltooth sawfish

Pristis pectinata Latham, 1794

DISTINGUISHING FEATURES: Elongate depressed fish with long narrow rostrum protruding from snout containing 24 to 32 pairs of teeth. Gill slits on ventral surface of head. Pectoral fins broad, small, and connected to head ahead of first gill slit, yet posterior to mouth. Two dorsal fins far back on body; caudal fin. Teeth 88-128/84-176.

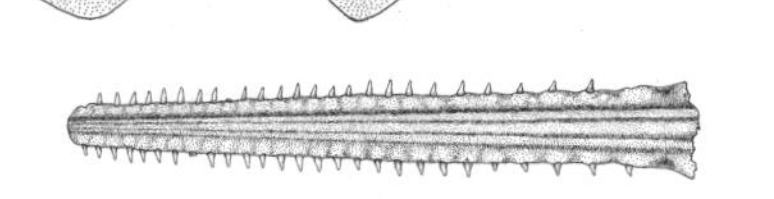

COLOR: Dark gray to brown, pale white below.

DISTRIBUTION AND OCCURRENCE: New Jersey to central Brazil, Gulf of Mexico. A near-shore species that will enter estuaries, sounds, and even fresh water. Five known captures in North Carolina prior to 1960. Largest captured June 1963 off Carolina Beach. A 1.2-meter smalltooth sawfish was gill netted and released in the "drain" of Barden's Inlet (Carteret County), North Carolina, 15 July 1999. Largest smalltooth sawfish from South Carolina was 4 meters long, captured off Georgetown, 7 August 1958. Occurrences were July–August in 10-meter water depths. Species now rare and endangered throughout its range.

SIZE: Attains 6.1-meter lengths, in both Carolinas to 4.8 meters, once in sounds and brackish waters.

REFERENCES: Adams and Wilson 1995; Bigelow and Schroeder 1953; Ishihara, Taniuchi, and Shizmizu 1991; Jenkins 1985; Jordan and Gilbert 1882; Miller 1995; Radcliffe 1916; Schwartz 1984; Slaughter and Springer 1968; Smith 1907; Thorson 1973; Weitzman 1997; Yarrow 1877.

SNOUT V-SHAPED, TRANSLUCENT, ELONGATE, AND LACKS LATERAL TEETH

Atlantic guitarfish

Rhinobatos lentiginosus Garman, 1880

DISTINGUISHING FEATURES: Body flattened, long, triangular shaped. Eyes and spiracles on top of head. Posterior margin of spiracle with two ridges. Snout long, tip thin, translucent. Preoral length 3 times mouth width. Dorsal fins far back on body. First dorsal fin originates over pelvic fins. Tail lacks spines. Distinct lower caudal fin lobe. Teeth 56-80/51-82.

COLOR: Gray, tan to chocolate brown dorsally, pale white ventrally. May have small white spots scattered over body.

DISTRIBUTION AND OCCURRENCE: Cape Lookout, North Carolina, to Florida to Yucatan, Mexico. Rare north of Cape Hatteras. Inshore species March–November; offshore year-round in deeper shelf waters. Known in Port Royal Sound, South Carolina.

SIZE: Elsewhere attains 75-centimeter lengths. North Carolina specimens attain 90-centimeter lengths.

REFERENCES: Bigelow and Schroeder 1953, 1962; Jordan and Gilbert 1882; Radcliff 1916; Schwartz 1984, 1995; Smith 1907; Weitzman 1997.

Deep-sea blind ray

Benthobatis marcida Bean and Weed, 1909

DISTINGUISHING FEATURES: Oval-shaped body disc. Possesses electric organs in dorsal surface of disc. Tail long. Eyes minute and covered with skin. Pelvic fins originate slightly anterior to axil of pectoral fin and overlap rear posterior margin of pectoral fin. Inner edge of pelvic fins joined along entire length to tail; lacks lateral membranes along tail. Two similar-sized dorsal fins. Skin flabby; caudal fin ovate.

COLOR: Light tan to brown-whitish below.

DISTRIBUTION AND OCCURRENCE: North Carolina to Gulf of Mexico in waters to 920 meters deep. Year-round off Carolinas in waters to 270–907 meters deep. Two specimens captured in 1909 were in 115–122-meter waters; one 29 July 1970 in deep water.

SIZE: Attains sizes of 0.5 meters.

REFERENCES: Bearden 1965a, 1965b; Bigelow and Schroeder 1953; Clark and Kristoff 1990; Daiber 1959; de Carvalho 1999; McEachran and Fechhelm 1998; Springer and Bullis 1956; Weitzman 1997.

Lesser electric ray

Narcine brasiliensis (Olfers, 1831)

DISTINGUISHING FEATURES: Short body with round body disc, stout tail. Two dorsal fins tall. Two large elongate oval electric organs visible on top of disc from front of eyes laterally to rear end of disc; shock powerful. Spiracles behind eyes on top of head. Gill slits small. Pelvic fins originate at axils of pectoral fins. First dorsal fin begins posterior to ends of pelvic fins. Caudal fin an equilateral triangle shape. Membranes along lateral sides of tail. Teeth 17-34/17-34.

COLOR: Light tan to brown dorsally; white ventrally. Posterior ventral edges of disc and pelvic fins dusky. Large dusky to black rings or loops on dorsal surface of disc with light centers.

DISTRIBUTION AND OCCURRENCE: North Carolina to southern Brazil, and occasionally off northern Argentina. Occurs offshore year-round in Carolinas in waters to 108 meters deep. Frequents inshore areas April–May in waters 5 meters deep, June–October in shelf waters. Will enter estuaries. Early published reports often confused this ray with stargazers, *Astroscopus* spp., which also have electric organs on the top of their bulky heads.

SIZE: Attains 450-millimeter lengths.

REFERENCES: Bearden 1965a, 1965b; Bigelow and Schroeder 1953, 1962; Gudger 1913a; Radcliffe 1916; Rudloe 1989; Schwartz 1984, 1989b, 1995, 1999, 2000a; Schwartz, Hogarth, and Weinstein 1982; Weitzman 1997.

Atlantic torpedo

Torpedo nobiliana Bonaparte, 1835

DISTINGUISHING FEATURES: Large round disc with straight anterior margin. Disc wider than long. Eyes and spiracles on top of head. Possesses powerful elongate oval electric organs on each side of dorsal disc from eyes to near end of disc. Gill slits small. Pelvic fins originate at axil of pectoral fin and are slightly overlapped by rear of pectoral fins. Pelvic fins far from tail. Origin of first dorsal fin is posterior to axils of pelvic fins. Second dorsal fin base half that of first dorsal fin base. Caudal fin triangular shaped. Lateral folds on tail. No spines on tail. Teeth 38-66/38-66.

COLOR: Jet black dorsally, no spots dorsally; white ventrally, margins dusky black.

DISTRIBUTION AND OCCURRENCE: Nova Scotia to North Carolina, perhaps Florida Keys; Gulf of Mexico to Venezuela in waters to 530 meters deep, but usually shallower. North Carolina inshore in April–May near Cape Lookout and Bogue Banks; will enter estuaries. Occasionally captured in South Carolina.

SIZE: Attains lengths of 1.8 meters. North Carolina specimens 2 meters long by 1.5 meters wide.

REFERENCES: Bearden 1961, 1965a; Bigelow and Schroeder 1953, 1962; Jordan and Gilbert 1882; Schwartz 1984; Weitzman 1997.

TWO DORSAL FINS, WELL-DEVELOPED CAUDAL FIN, TAIL NOT WHIPLIKE, LATERAL TIPS OF PECTORAL FIN DISC ELONGATE AND POINT FORWARD

Skilletskate

Dactylobatus armatus Bean and Weed, 1909

DISTINGUISHING FEATURES: Body round. Lateral edges midway on round body; disc pectoral fins elongate and pointed forward. Two dorsal fins near end of tail confluent at base; no thorn between them. Tail with narrow lateral folds. Males possess sharp thorns around edge of disc ventrally to pectoral forward lobe extensions. No thorns on pelvic fins. Eyes and spiracles on top of head. No electric organs. Teeth 54-66/66.

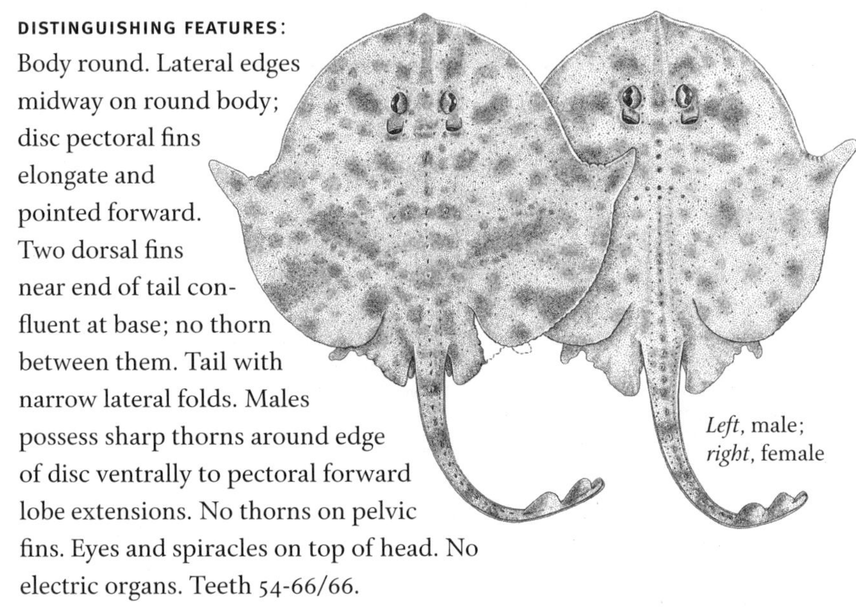

Left, male; *right*, female

COLOR: Brown-gray above with black dusky spots scattered on disc and caudal fin. White dusky below.

DISTRIBUTION AND OCCURRENCE: South Carolina to Gulf of Mexico and off Nicaragua to Venezuela in waters to 685 meters deep. Several specimens captured in 1880s off Charleston, South Carolina, to Cape Kennedy, Florida. Largest recent North Carolina capture, female 662 millimeters total length in 464-meter waters. Recently rare.

SIZE: Usually attains 316-millimeter total lengths.

REFERENCES: Bean and Weed 1909; Bearden 1965a; Bigelow and Schroeder 1953, 1965, 1968b; Blackman 1972; McEachran and Fechhelm 1998; Schwartz 1984.

Caribbean skate

Dipturus teevani (Bigelow and Schroeder, 1951)

DISTINGUISHING FEATURES: Snout long and sharp. Greatest wing width behind midline of body. Snout-mouth distance 19 to 25 percent of total length. Disc width 73 to 81 percent of total length. Two dorsal fins separated. Eyes and spiracles on dorsal surface of head. Tail with lateral folds. Tail widest rearward of dorsal fins; on all other skates it is narrower. Previously known as *Raja floridana*. Teeth 36-38/36-38.

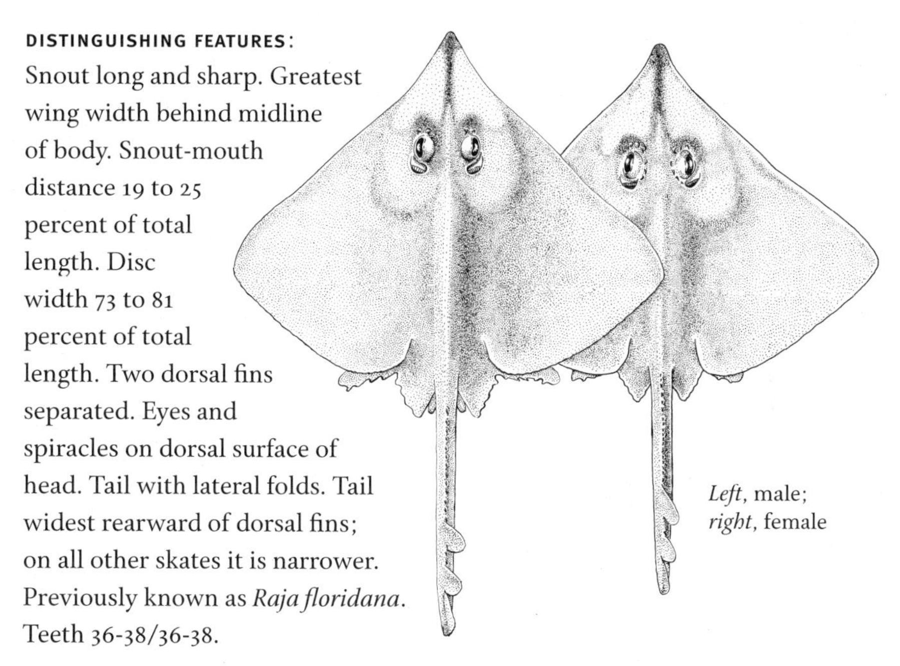

Left, male; *right*, female

COLOR: Brown above, no spots; black on dorsal and caudal fins. White below. Mucous pores in ventral surface anterior to mouth.

DISTRIBUTION AND OCCURRENCE: North Carolina, Gulf of Mexico to Caribbean to Colombia, South America, in waters to 732 meters deep. Occurs from Cape Lookout, North Carolina, southward to Caribbean May–June in 306- to 405-meter waters.

SIZE: Attains 840-millimeter total lengths; in Carolinas, 400 millimeters total length.

REFERENCES: Bearden 1965a; Bigelow and Schroeder 1951, 1953, 1962, 1965, 1968a; Clark and Kristoff 1990; Jacob and McEachran 1994; McEachran and Fechhelm 1998; Schwartz 1984.

Barndoor skate

Dipturus laevis (Mitchill, 1817)

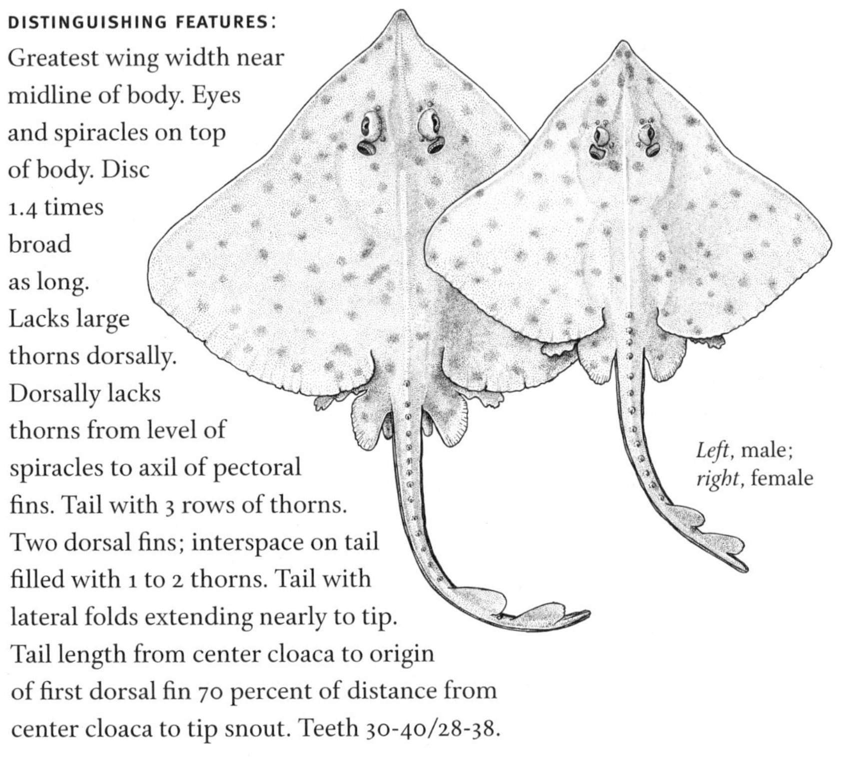

Left, male; *right*, female

DISTINGUISHING FEATURES: Greatest wing width near midline of body. Eyes and spiracles on top of body. Disc 1.4 times broad as long. Lacks large thorns dorsally. Dorsally lacks thorns from level of spiracles to axil of pectoral fins. Tail with 3 rows of thorns. Two dorsal fins; interspace on tail filled with 1 to 2 thorns. Tail with lateral folds extending nearly to tip. Tail length from center cloaca to origin of first dorsal fin 70 percent of distance from center cloaca to tip snout. Teeth 30-40/28-38.

COLOR: Brown above with scattered darker spots of various sizes. White ventrally; may be blotched. Mucous pores anterior of mouth as black dots or dashes.

DISTRIBUTION AND OCCURRENCE: Nova Scotia to Cape Lookout, North Carolina, and rarely South Carolina in waters to 434 meters deep. Recorded from South Carolina as *Raja floridana*. Moves over shelf as waters warm and cool.

SIZE: One meter.

REFERENCES: Bearden 1965a; Bigelow and Schroeder 1953, 1954; Radcliffe 1916; Smith 1907.

DORSAL FINS CONFLUENT, DISC COLORED DORSALLY AND VENTRALLY

Deepwater skate

Rajella bathyphila (Holt and Byrne, 1908)

DISTINGUISHING FEATURES: Eyes and spiracles dorsally on head. Large thorns near snout. Two dorsal fins confluent except in mature fish. Disc nearly perfect diamond, 1.1 times broad as long. Tail with narrow lateral folds along two-fifths of its length. Long tail, 1.5 times distance from cloaca to tip of tail as cloaca to tip of snout. Small prickles on tail. Teeth 37 or 39/37 or 39.

Left, male; *right*, female

COLOR: Brown-gray above. Ventrally chocolate brown on disk and near pelvic edges.

DISTRIBUTION AND OCCURRENCE: Nova Scotia to North Carolina in 2,385-meter waters. Rare in North Carolina to 2,173-meter waters in summer.

SIZE: Attains 463-millimeter lengths.

REFERENCES: Bigelow and Schroeder 1953, 1962; McEachran and Fechhelm 1998; McEachran and Miyake 1990a, 1990b; McEachran and Stehmann 1977; Schwartz 1984; Stehmann 1986.

Richardson's skate

Bathyraja richardsoni (Garrick, 1961)

DISTINGUISHING FEATURES: Disc nearly diamond shaped, broader than long. Eyes and spiracles on top of head. Tail short with narrow ventrolateral fold on each side from pelvic fin base to middle of second dorsal fin base. Tail length from center of cloaca to first dorsal fin origin 50 percent of distance from cloaca to tip of snout. Interspace between dorsal fins 4 times the length of first dorsal fin base. Lacks large thorns on tail. Teeth 14-1-16/14-1-13.

COLOR: Light gray-brown above, ventrally darker gray with creamy blotches on pectoral fin margins and lower surface of disc and tail. White around mouth, nasal flap, nostrils, gill slits, and cloaca. White inside mouth.

Top, male; *bottom*, female

DISTRIBUTION AND OCCURRENCE: Nova Scotia to North Carolina and elsewhere to 2,360-meter depths.

SIZE: Attains 1.7-meter total length; 80-millimeter disk width.

REFERENCES: Bigelow and Schroeder 1950; Garrick 1961; McEachran and Miyake 1984; Stehmann 1986; Templeman 1973.

Pluto skate

Fenestraja plutonia (Garman, 1881)

DISTINGUISHING FEATURES: Disc 1.2 times wide as long; pear shaped. Eyes and spiracles on top of head. Slender tail; two dorsal fins confluent, separate in *F. atripinna*. Tail with lateral folds. Distance from tip of tail to pectoral fin 1.4 to 1.7 times distance from center of cloaca to tip of snout. Five to 7 bars on tail. Tail long. Previously placed in genera *Breviraja* and *Gurgesiella*. Teeth 32-34/32-34.

COLOR: Brown-gray, dense spots or blotches on dorsal surface. Yellow-white below.

DISTRIBUTION AND OCCURRENCE: North Carolina to Venezuela and Guyanas in 1,024-meter waters; 732-meter waters in North Carolina. Rare but known from off Cape Lookout, North Carolina, in winter in 463- to 732-meter waters.

Top, male; *bottom*, female

SIZE: Maximum size 270 millimeters total length.

REFERENCES: Bigelow and Schroeder 1953, 1962, 1968a; McEachran and Fechhelm 1998, Schwartz 1984.

DORSAL FINS CONFLUENT, DORSALLY THREE OR MORE ROWS OF THORNS ALONG MIDDLE OF DISC

Spinose skate

Breviraja spinosa Bigelow and Schroeder, 1950

DISTINGUISHING FEATURES: Obtuse snout, 145° angle in front of spiracles. Three rows of thorns along each side of midline of disc. Mouth arched on each side of symphysis. Dorsal fins confluent; second dorsal fin similar to first. Tail about 60 percent of total length; has lateral folds on posterior third of length. Ventral surface naked. Teeth 40-44/36-44.

COLOR: Tan, no spots or marks dorsally, white ventrally; occasionally dusky gray area near center of disc ventrally.

DISTRIBUTION AND OCCURRENCE: North Carolina to Florida Keys, Gulf of Mexico in 620-meter waters; off North Carolina in 450- to 549-meter waters.

SIZE: Attains 330-millimeter total lengths.

REFERENCES: Bigelow and Schroeder 1950, 1953, 1962; McEachran and Compagno 1982; McEachran and Matheson 1985; Schwartz 1984.

Top, male; *bottom*, female

Winter skate

Leucoraja ocellata (Mitchill, 1815)

DISTINGUISHING FEATURES: Similar to little skate. Broad pear-shaped disc. Disc 1.3 times broad as long. Snout angle 145° in front of spiracles. Distance between orbits less than 12 percent of total length. Clear area either side of snout. Young difficult to distinguish from *Leucoraja erinacea*. Tail narrow; lateral folds from near beginning of tail to its tip. Tail length from cloaca to tip of snout 1.1 times as great as distance from cloaca to snout. Two confluent dorsal fins similar size and shape. Teeth 72-110/74-100.

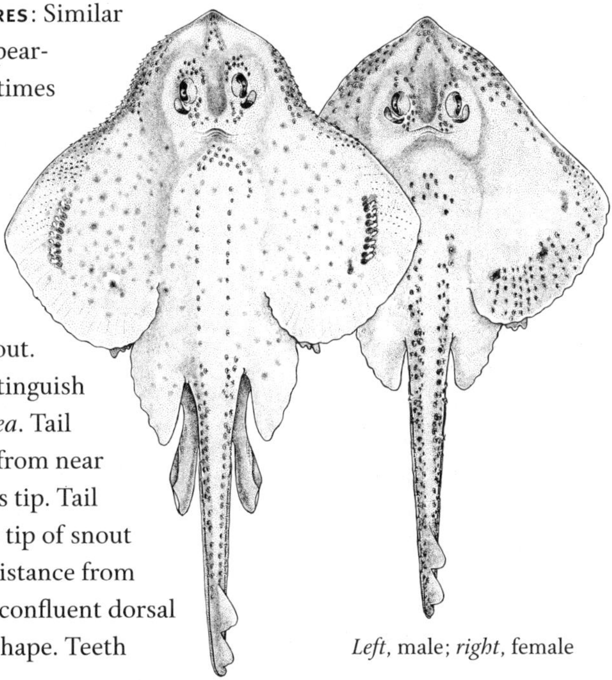

Left, male; *right*, female

COLOR: Light brown above with round, scattered, blackish spots on disc, pelvic fins, and tail. Ventrally white with pale, irregular-sized blotches on rear of disc and tail.

DISTRIBUTION AND OCCURRENCE: Nova Scotia to North Carolina in 205-meter waters off North Carolina near Cape Hatteras or in Albemarle Sound in winter.

SIZE: Attains 806-millimeter lengths.

REFERENCES: Bigelow and Schroeder 1953, 1962; McEachran and Martin 1978; McEachran and Musick 1973, 1975; Schwartz 1984; Templeman 1965.

Little skate

Leucoraja erinacea (Mitchill, 1825)

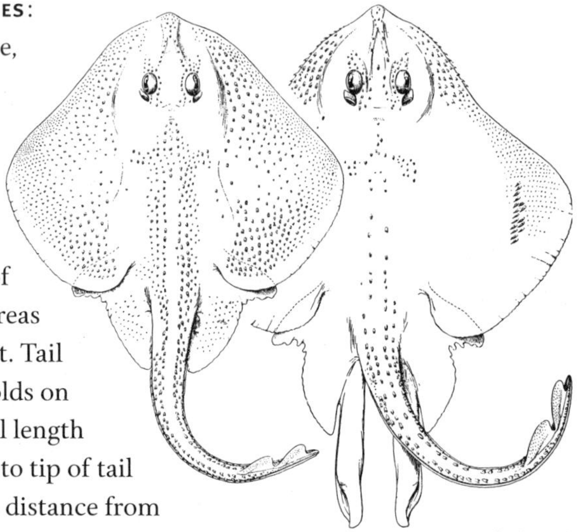

DISTINGUISHING FEATURES: Similar to winter skate, but distance between orbits more than 12 percent of total length. Disc 1.2 times broad as long. Snout angle in front of spiracles 125°. Clear areas on either side of snout. Tail with narrow lateral folds on outer third of tail. Tail length from center of cloaca to tip of tail is 1.3 times as great as distance from center of cloaca to snout. Midline row of large thorns on disc extending onto tail. Two similar-sized and -shaped dorsal fins. Teeth 38-64/38-64.

Left, female; *right*, male

COLOR: Dark brown with some dark spots on disc. Ventrally white; tail lower surface dark gray.

DISTRIBUTION AND OCCURRENCE: Nova Scotia to North Carolina to 144-meter waters. Rarely reaches North Carolina in 199-meter waters in winter.

SIZE: Attains 530-millimeter lengths.

REFERENCES: Bigelow and Schroeder 1953, 1962; McEachran and Martin 1978; McEachran and Musick 1973; Schwartz 1984; Templeman 1965.

Smooth skate

Malacoraja senta (Garman, 1885)

DISTINGUISHING FEATURES: Snout angle in front of spiracles 110°. Two similar-sized dorsal fins confluent. Midline of thorns diminish in size posterior of pectoral fins. Narrow tail folds along posterior two-thirds of tail. No prickles on center portion of spade-shaped disc. Disc 1.25 times broad as long. Tail length from center of cloaca to first dorsal fin about 85 percent of distance from center of cloaca to tip of snout. Teeth 38-40/36-38.

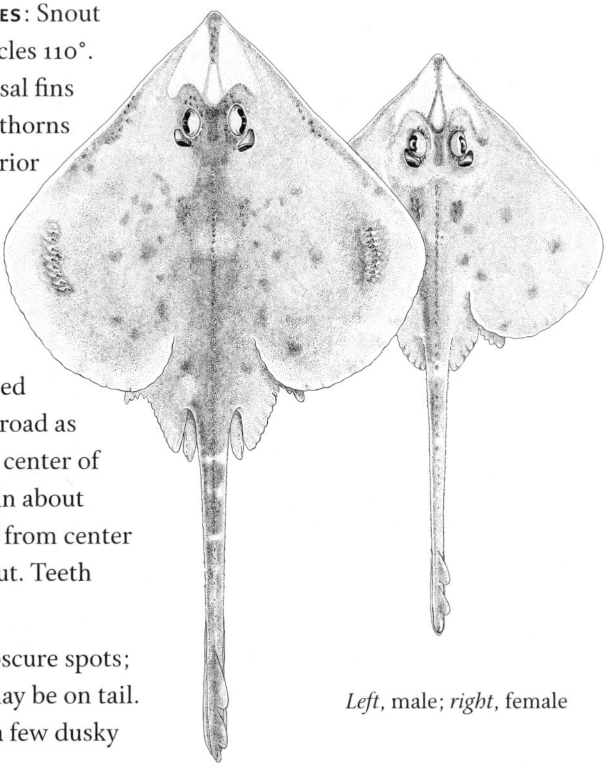

COLOR: Brown with obscure spots; two faint cross bars may be on tail. Ventrally white with a few dusky blotches.

Left, male; *right*, female

DISTRIBUTION AND OCCURRENCE: Nova Scotia to Charleston, South Carolina, in 956-meter waters. A rare northern species in the Carolinas.

SIZE: Attains 500-millimeter lengths.

REFERENCES: Bearden 1965a; Bigelow and Schroeder 1953, 1962; Radcliffe 1916; Schwartz 1984.

Carolina pygmy skate

Neoraja carolinensis McEachran and Stehmann, 1984

DISTINGUISHING FEATURES: Snout short. Disc 1.2 times broad as long. Snout angle 120° at level of spiracles. Distance between orbits 3.8 to 4.3 percent of total length. Tail slender with lateral folds on last half. Distance from tail to center of cloaca is 1.4 times distance from tip of snout to center of cloaca. Dorsal fins same size and confluent. No crossbars on tail. Teeth 40-44/40-44.

COLOR: Gray-brown dorsally, white ventrally except dusky black on abdomen and blotched elsewhere.

Top, male; *bottom*, female

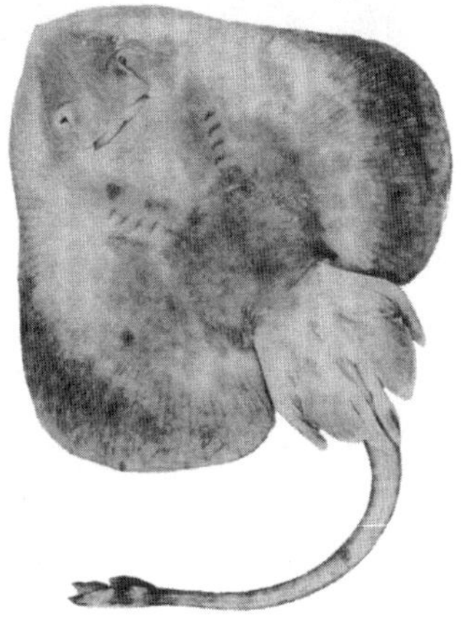

DISTRIBUTION AND OCCURRENCE: Off Cape Lookout, Cape Fear River area North Carolina to Florida. Four young specimens collected between 33° and 34°N, 75 and 76′W in 695- to 1,010-meter waters, February, June 1972 and 1979.

SIZE: Attains total lengths of 143 to 285 millimeters.

REFERENCES: McEachran and Compagno 1982; McEachran and Stehmann 1984.

DORSAL FINS SEPARATED, WITH OR WITHOUT INTERSPACE THORN

Blackfin pygmy skate

Fenestraja atripinna (Bigelow and Schroeder, 1950)

DISTINGUISHING FEATURES: Disc 1.2 times broad as long. Snout length twice distance in front of orbits. Snout pointed in *F. atripinna*, broad in *F. plutonia*. Nubbin at tip of snout. Two dorsal fins wide apart, confluent in *F. plutonia*. Distance between dorsal fins 40 percent of first dorsal fin length. Tail long. Teeth 40/40.

COLOR: Plain brown without black markings. No bars on tail. Dorsal and caudal fins black. White ventrally, some chocolate brown near pelvic fins.

DISTRIBUTION AND OCCURRENCE: Labrador south to Charleston, South Carolina, and Cuba in 252-meter waters. Off Cape Fear River in deep shelf waters.

SIZE: Attains 700-millimeter lengths; 530-millimeter disc widths.

REFERENCES: Bigelow and Schroeder 1953, 1962, Schwartz 1984.

Thorny skate

Amblyraja radiata (Donovan, 1808)

DISTINGUISHING FEATURES: Snout in front of orbits 2.5 times distance between orbits. Disc thorny or prickly. Large thorns on midline of disc from nape to first dorsal fin, 10 larger than on tail. Disc diamond shaped. Interspace between dorsal fins. Tail 80 percent of distance from cloaca to snout. Tail with lateral folds from axil of pelvic fin to end. Teeth 36-46/36-46.

COLOR: Brown dorsally; may have white spot beside eye. Ventrally white with some dusky smudges.

DISTRIBUTION AND OCCURRENCE: Labrador south to Charleston, South Carolina, in 97-meter waters.

SIZE: Attains 1,020-millimeter lengths.

REFERENCES: Bigelow and Schroeder 1953, 1962; Schwartz 1984.

Rosette skate

Leucoraja garmani (Whitley, 1939)

DISTINGUISHING FEATURES: Once recognized as four subgenera, now as four distinct species. Snout short, 10 percent of total length; maximum angle about 120° before spiracles. Mouth strongly arched. Two dorsal fins; interspace 10 to 18 percent of first dorsal fin base length. Tail 60 percent of total length. Lateral folds extend most of tail length. Thorns in 2 to 5 rows on body to level of second dorsal fin. Once known as *Raja lentignosa*. Teeth 46-55/46-55.

Left, male; *right*, female

COLOR: Brown dorsally, with dark spots resembling rosettes on dorsal disc; white below. Dark spots on tail form bars.

DISTRIBUTION AND OCCURRENCE: Cape Hatteras to Dry Tortugas off edge of continental shelf in waters to 366 meters deep.

SIZE: Attains 335-millimeter lengths.

REFERENCES: Bearden 1965a; Bigelow and Schroeder 1951, 1953, 1954, 1962, 1965, 1968a; McEachran 1970, 1977b; McEachran and Compagno 1982; McEachran and Fechhelm 1998; McEachran and Musick 1975; Radcliffe 1916; Schwartz 1984.

Clearnose skate

Raja eglanteria Bosc, 1802

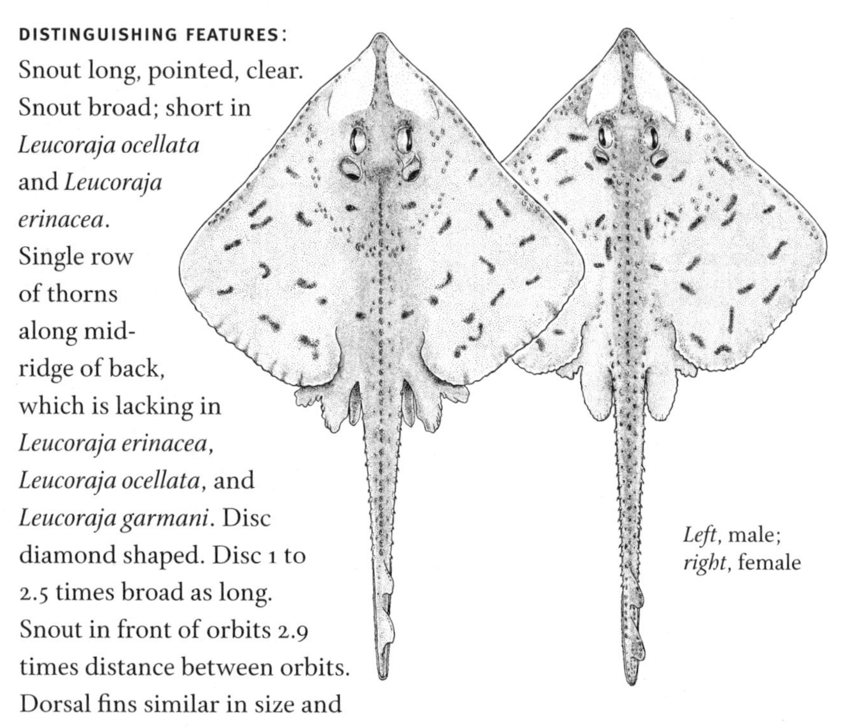

Left, male; *right*, female

DISTINGUISHING FEATURES: Snout long, pointed, clear. Snout broad; short in *Leucoraja ocellata* and *Leucoraja erinacea*. Single row of thorns along mid-ridge of back, which is lacking in *Leucoraja erinacea*, *Leucoraja ocellata*, and *Leucoraja garmani*. Disc diamond shaped. Disc 1 to 2.5 times broad as long. Snout in front of orbits 2.9 times distance between orbits. Dorsal fins similar in size and shape; interspace 14 to 30 percent of first dorsal fin base. Tail with lateral folds. Tail 49 to 52 percent of total length. Teeth 46-54/48-49.

COLOR: Tan-brown dorsally; bars, dashes, or blotches scattered over surface of disc dorsally; white below.

DISTRIBUTION AND OCCURRENCE: Massachusetts to Florida, Gulf of Mexico in waters to 117 meters deep. Common year-round in Carolinas, inshore in sounds and rivers February–November. Moves offshore when waters warm above 12°C.

SIZE: Attains 785-millimeter lengths.

REFERENCES: Bearden 1965a; Bigelow and Schroeder 1953, 1962; McEachran and Fechhelm 1998; McEachran and Musick 1975; Radcliffe 1916; Schwartz 1995, 1996a, 2000a; Schwartz, Hogarth, and Weinstein 1982; Smith 1907.

NO DORSAL FIN, TAIL WELL DEVELOPED, SPINE AT CAUDAL FIN, BOTH NEAR END OF TAIL

Yellow stingray

Urobatis jamaicensis (Cuvier, 1817)

DISTINGUISHING FEATURES: Spiracles and eyes on top of head. No dorsal fin. Disc round, 0.9 times broad as long. Caudal fin not whiplike. Serrated spine near end of tail near caudal fin. Tail 47 percent of total length. Tail has low keel on each side. Teeth 30-34/28-30.

COLOR: Gray-greenish above with white, yellow, gold spots scattered on a dark background dorsally; white ventrally and ventral surface of tail.

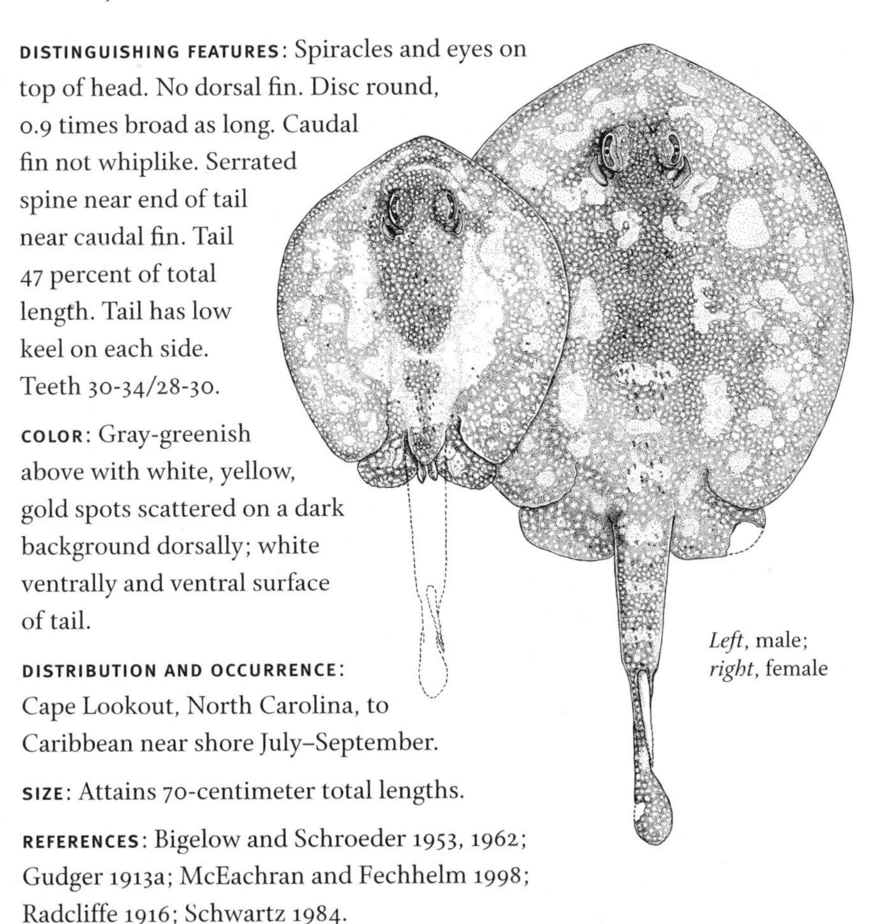

Left, male; *right*, female

DISTRIBUTION AND OCCURRENCE: Cape Lookout, North Carolina, to Caribbean near shore July–September.

SIZE: Attains 70-centimeter total lengths.

REFERENCES: Bigelow and Schroeder 1953, 1962; Gudger 1913a; McEachran and Fechhelm 1998; Radcliffe 1916; Schwartz 1984.

Pelagic stingray

Pteroplatytrygon violacea (Bonaparte, 1832)

DISTINGUISHING FEATURES: Disc widest near or before midline through level of eyes. Disc 1.3 times broad as long. Anterior margin of disc forms shallow arc, not U-shape. Caudal fin with serrated spine(s). No dorsal ridge or fold on tail. Ventral fin fold extends from spine tip to tip of tail. Marked sexual dimorphism in teeth. Previously in genus *Dasyatis*. Teeth 28-34 in upper jaw.

COLOR: Jet black-purple all over. Naked dorsally and ventrally.

DISTRIBUTION AND OCCURRENCE: From Greenland to North Carolina, Gulf of Mexico to Texas, and Lesser Antilles, to southern Brazil, worldwide. At edge of continental shelf in open ocean.

SIZE: Attains 800-millimeter disc width.

REFERENCES: Bigelow and Schroeder 1953, 1962; Branstetter and McEachran 1983; McEachran and Fechhelm 1998; Menni and Stehmann 2000; Scott and Tibbo 1968; Wilson and Beckett 1970.

Atlantic stingray

Dasyatis sabina (Lesueur, 1824)

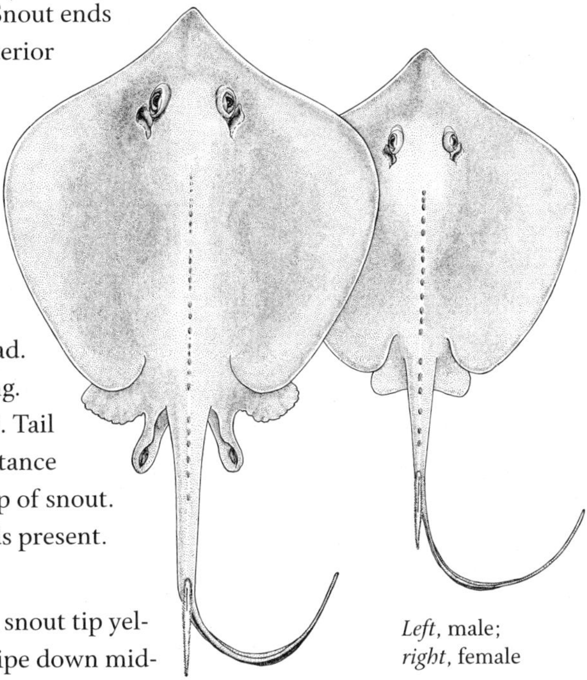

DISTINGUISHING FEATURES: Snout ends in elongated point and anterior to eyes is longer than distance between spiracles; snout more triangular shaped. Anterior angle of disc is 53° to level between eyes. Eyes and spiracles on top of head. Disc 1.1 times broad as long. Tail with serrated spine(s). Tail length twice as long as distance from center of cloaca to tip of snout. Dorsal and ventral fin folds present. Teeth 28-26/28-36.

Left, male; *right,* female

COLOR: Light tan disc with snout tip yellowish. May have dark stripe down midline of back. Ventrally white; lateral edge of disc and tail yellowish.

DISTRIBUTION AND OCCURRENCE: Chesapeake Bay to Gulf of Mexico. Very common in Carolinas, even resident in South Carolina. Enters sounds, estuaries, and rivers and usually occurs in shallow ocean waters to 36 meters deep. February–November migrates along the shore seasonally.

SIZE: Attains 610-millimeter disc widths.

REFERENCES: Bearden 1965a; Bigelow and Schroeder 1953, 1962; Jordan and Gilbert 1882; Kajura and Tricas 1996; Radcliffe 1913, 1916; Schwartz 1984, 1995, 2000a; Schwartz and Dahlberg 1978; Teaf and Lewis 1987.

Bluntnose stingray

Dasyatis say (Lesueur, 1817)

DISTINGUISHING FEATURES: Snout anterior to eyes shorter than distance between spiracles. Angle of anterior disc at snout 59°. Snout not very pointed. Snout before eyes 0.7 to 0.8 times outer orbit distance. Tips of wings rounded. Few small thorns at midline behind spiracles, more on tail. No caudal fin but has serrated spine on tail. Black fin folds on tail. Teeth 36-56/36-56.

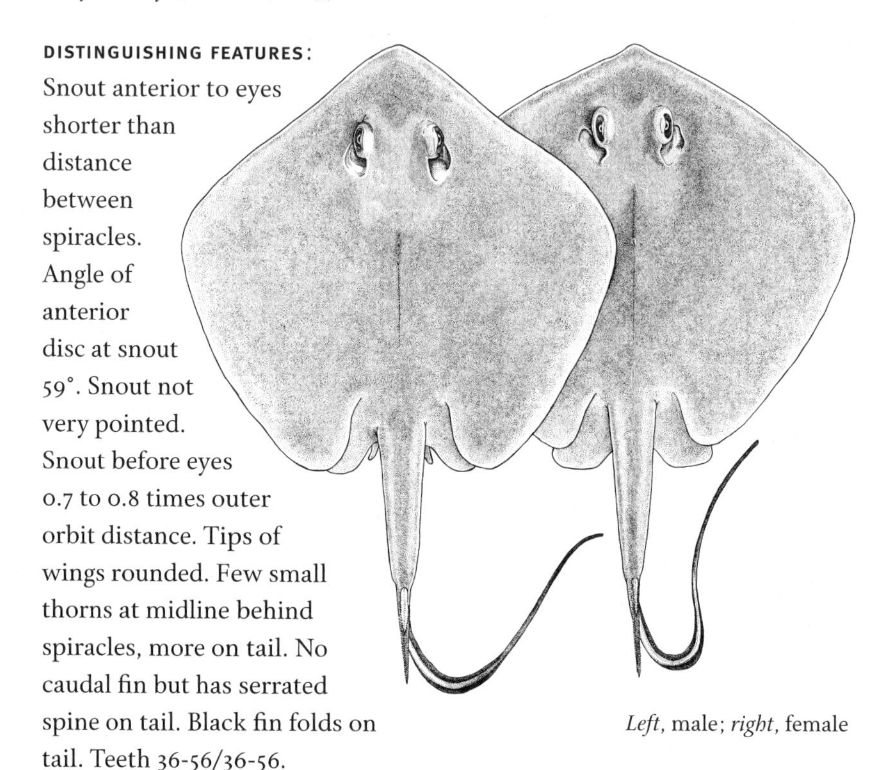

Left, male; *right*, female

COLOR: Brown dorsally, white ventrally.

DISTRIBUTION AND OCCURRENCE: Massachusetts to southern Brazil, Gulf of Mexico in waters to 9 meters deep. March–November transient in Carolinas. Enters sounds, estuaries, and rivers.

SIZE: Attains 100-millimeter disc widths.

REFERENCES: Bearden 1961, 1965a; Bigelow and Schroeder 1953, 1962; Jenkins 1887; Radcliffe 1916; Schwartz 2000a; Schwartz, Hogarth, and Weinstein 1982; Smith 1907; Yarrow 1877.

Southern stingray

Dasyatis americana Hildebrand and Schroeder, 1928

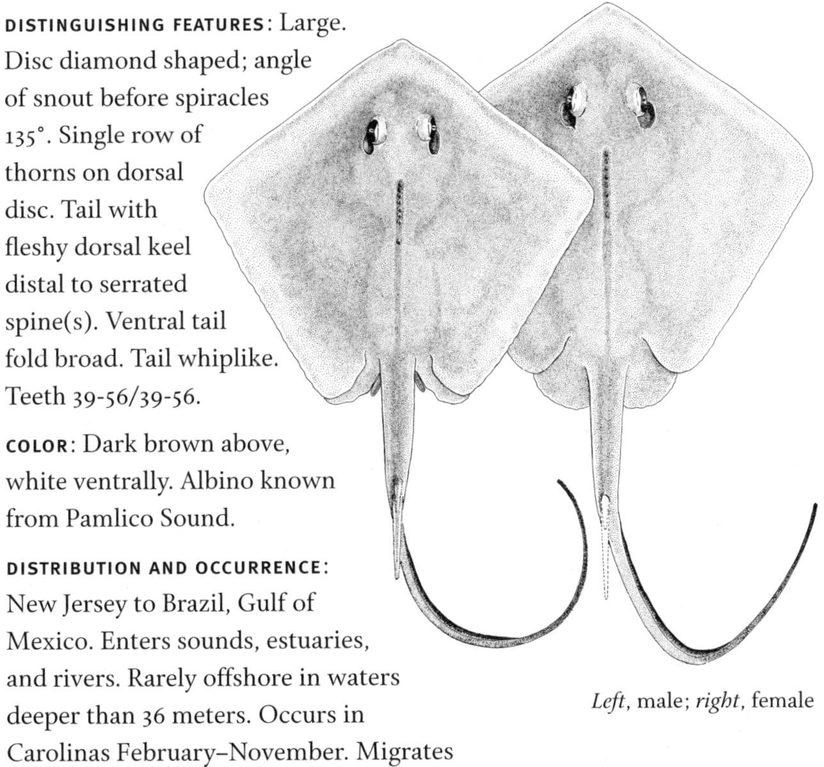

Left, male; *right*, female

DISTINGUISHING FEATURES: Large. Disc diamond shaped; angle of snout before spiracles 135°. Single row of thorns on dorsal disc. Tail with fleshy dorsal keel distal to serrated spine(s). Ventral tail fold broad. Tail whiplike. Teeth 39-56/39-56.

COLOR: Dark brown above, white ventrally. Albino known from Pamlico Sound.

DISTRIBUTION AND OCCURRENCE: New Jersey to Brazil, Gulf of Mexico. Enters sounds, estuaries, and rivers. Rarely offshore in waters deeper than 36 meters. Occurs in Carolinas February–November. Migrates north or south depending on water temperatures above 10° centigrade.

SIZE: Attains 1,100-millimeter disc widths. Will tolerate 0 parts per thousand salinities.

REFERENCES: Bearden 1961; Bigelow and Schroeder 1953, 1962; Gudger 1913a; Hildebrand and Schroeder 1928; Radcliffe 1916; Schwartz 1984, 1995, 2000a; Schwartz and Safrit 1977; Schwartz, Hogarth, and Weinstein 1982.

Roughtail stingray

Dasyatis centroura (Mitchill, 1815)

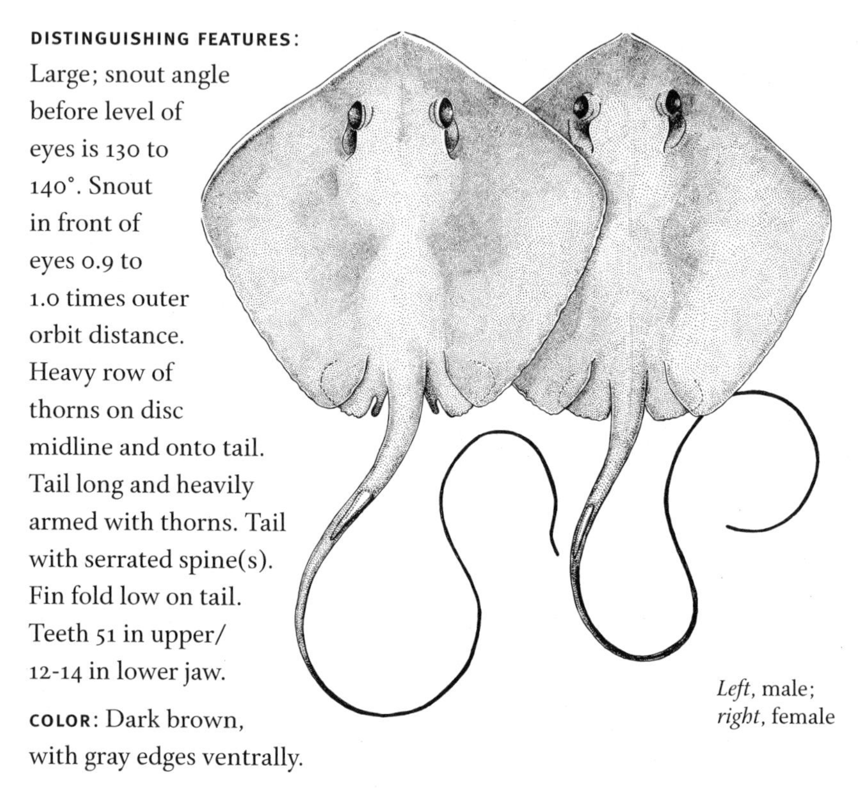

DISTINGUISHING FEATURES: Large; snout angle before level of eyes is 130 to 140°. Snout in front of eyes 0.9 to 1.0 times outer orbit distance. Heavy row of thorns on disc midline and onto tail. Tail long and heavily armed with thorns. Tail with serrated spine(s). Fin fold low on tail. Teeth 51 in upper/ 12-14 in lower jaw.

Left, male; *right*, female

COLOR: Dark brown, with gray edges ventrally.

DISTRIBUTION AND OCCURRENCE: Massachusetts to Uruguay in waters to 54 meters deep. Occurs April–October in Carolinas; will enter estuaries, sounds, and rivers.

SIZE: Attains 210-centimeter disc widths.

REFERENCES: Bearden 1965a, 1965b; Bigelow and Schroeder 1953, 1961; Bullis and Struhsaker 1961; Schwartz 1995, 2000a; Schwartz, Hogarth, and Weinstein 1982.

NO DORSAL FIN, SHORT WHIPLIKE TAIL, BODY BROADER THAN LONG

Spiny butterfly ray

Gymnura altavela (Linnaeus, 1758)

DISTINGUISHING FEATURES: Body naked. Disc margin angle to 135°. Disc broader than long; tentacle projection in posterior side of spiracle. Serrated spine(s) on short tail. Tail keeled with dorsal and ventral keels. Tail one-fourth of disc width. Teeth 98-138/78-110.

COLOR: Olive green disc dorsally. Disc may be edged with cream-colored spots. Dorsal disc and tail may have some white spots. Ventrally white.

DISTRIBUTION AND OCCURRENCE: Massachusetts to Rio de la Plata, Argentina. Common, occurs in Carolinas February–October in waters to 54 meters deep. Will enter estuaries.

SIZE: Attains 2.6-meter widths.

REFERENCES: Bigelow and Schroeder 1953, 1962; Coles 1915; Gudger 1913b; Jenkins 1887; Radcliffe 1916; Schwartz 1995, 2000a; Schwartz, Hogarth, and Weinstein 1982; Wilson 1900; Yarrow 1877.

Smooth butterfly ray

Gymnura micrura (Bloch and Schneider, 1801)

DISTINGUISHING FEATURES: Disc broader than long. No tentacle in spiracle. Anterior snout margin disc angle 130°. Tail lacks serrated spines. Tail has dorsal and ventral ridges. Body naked. Earlier placed in genus *Pteroplatea*.

Left, male; *right*, female

COLOR: Gray–olive green dorsally. Edges of disk rimmed with cream-colored spots. Upper disc dotted with dark blotches. Ventrally white. Teeth 60-120/53-106.

DISTRIBUTION AND OCCURRENCE: Maryland to Brazil. Common April–November. Enter sounds, estuaries, rivers, and ocean to 10-meter depths.

SIZE: Attains 120-centimeter disc widths.

REFERENCES: Bearden 1961, 1965a; Bigelow and Schroeder 1953, 1962; Gudger 1910, 1912, 1913b; Radcliffe 1916; Schwartz 1984, 1995, 2000a; Schwartz, Hogarth, and Weinstein 1982; Smith 1907; Wilson 1900; Yarrow 1877.

EYES AND SPIRACLES ON SIDES OF HEAD,
HEAD WITH FORWARD EXTENSIONS

Manta ray

Manta birostris (Donndorff, 1798)

DISTINGUISHING FEATURES: Head preceded by curled cephalic fins. Mouth terminal at front edge of disc. Short tail with or without serrated spine. Body naked. Single dorsal fin at rear of disc. Teeth only in lower jaw about 270.

COLOR: Black-brown dorsally. Often has white collar across dorsal disc behind spiracles. White ventrally.

DISTRIBUTION AND OCCURRENCE: Worldwide. New England to Brazil, Gulf of Mexico, Bahamas, and Antilles. Occurs sporadically in Carolinas July–October; will enter inlets (Beaufort), sounds (Pamlico), and rivers (9 June 1971, Lockwood Folly River), and in waters to 7 meters deep. Known in North Carolina as early as 1709.

SIZE: Attains 6.7-meter disc widths. Most in North Carolina are 5 meters wide; one in Charleston, South Carolina, was 5.5 meters wide.

REFERENCES: Bearden 1961, 1965a; Bigelow and Schroeder 1953; Coles 1915, 1916; Gill 1903; Gudger 1912, 1913b; Ishihara, Honma, and Nakamura 2001; Jenkins 1887; Lawson 1709; Notobartolo-di-Sciara 1987; Notobartolo-di-Sciara and Hillyer 1989; Pearce and Williams 1951; Radcliffe 1916; Schwartz 1995; Smith 1907; Wilson 1900; Yarrow 1877.

Atlantic devil ray

Mobula hypostoma (Bancroft, 1831)

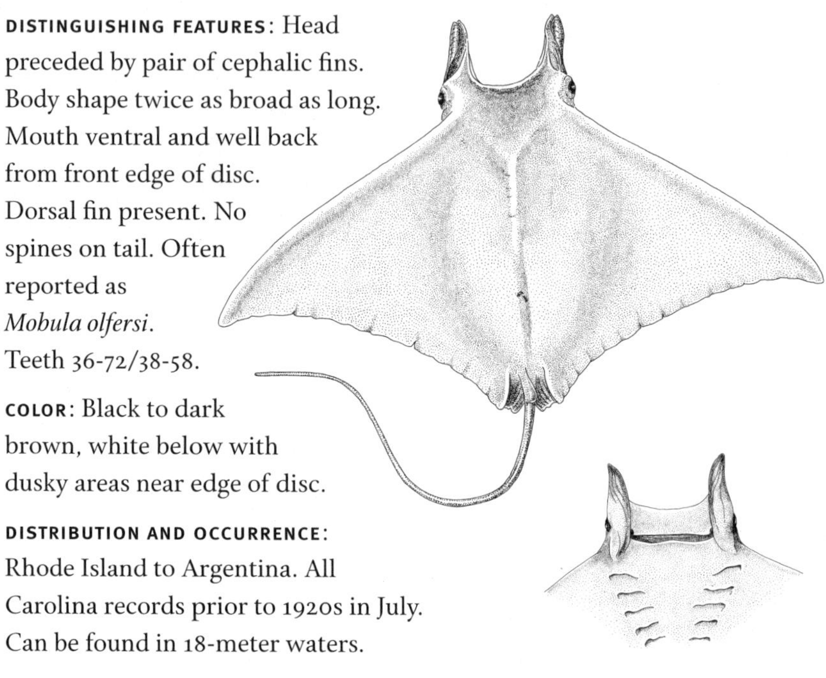

DISTINGUISHING FEATURES: Head preceded by pair of cephalic fins. Body shape twice as broad as long. Mouth ventral and well back from front edge of disc. Dorsal fin present. No spines on tail. Often reported as *Mobula olfersi*. Teeth 36-72/38-58.

COLOR: Black to dark brown, white below with dusky areas near edge of disc.

DISTRIBUTION AND OCCURRENCE: Rhode Island to Argentina. All Carolina records prior to 1920s in July. Can be found in 18-meter waters.

SIZE: Attains 1.3-meter disc widths.

REFERENCES: Bigelow and Schroeder 1953, 1962; Campbell and Monroe 1974; Coles 1910, 1913, 1915, 1916; Gill 1903; Gudger 1913a, 1913b; Notobartolo-di-Sciara 1987; Notobartolo-di-Sciara and Hillyer 1989; Radcliffe 1913, 1916.

Giant devil ray

Mobula mobular (Bonnaterre, 1788)

DISTINGUISHING FEATURES: Large size; head preceded by pair of cephalic fins. Mouth located well behind anterior edge of disc. Dorsal fin present. Tail with serrated spine(s). Teeth in each jaw 160 to 400, but number increases with specimen size. Use caution in distinguishing from *M. japonica*, which may be same as *M. mobular*.

COLOR: Black-brown dorsally, apex of dorsal fin white, white ventrally.

DISTRIBUTION AND OCCURRENCE: New Jersey to Cuba. Known in North Carolina from 3-meter-wide specimen caught August 1969 in 118-meter waters. No recent captures.

SIZE: North Carolina specimen 3.1 meters wide.

REFERENCES: Bigelow and Schroeder 1953, 1962; Notobartolo-di-Sciara 1987; Notobartolo-di-Sciara and Hillyer 1989; Schwartz 1984.

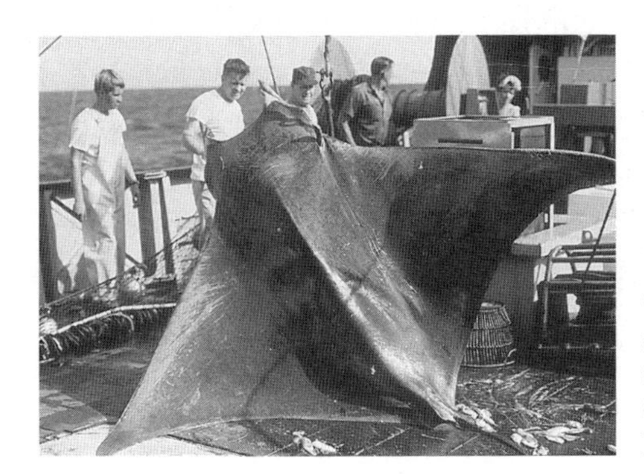

Giant devil ray, *Mobula mobular*, 3.1 meters wide, captured off the North Carolina coast in August 1969. Photo by J. Sterling.

Cownose ray

Rhinoptera bonasus (Mitchill, 1815)

DISTINGUISHING FEATURES: Head bilobed with fleshy flaps ventrally ahead of mouth. Dorsal fin present. Eyes and spiracles on side of head. Pavement teeth hexagonal in 6-11 series in each jaw. Body naked. Serrated spine(s) on whiplike tail. Teeth 7-11/7-11.

COLOR: Chocolate brown dorsally, white ventrally.

DISTRIBUTION AND OCCURRENCE: Atlantic population, Massachusetts to Rio de Janeiro; another population in Gulf of Mexico migrates to northern South America. Atlantic population migrates north and south seasonally, north by April 15 and southward by 30 October. Enters sounds, estuaries, and rivers. Young born in estuaries in June. Rarely moves into deep ocean waters, although can be found in 120-meter depths off Florida in late winter, and 460-meter depths in Gulf of Mexico (G. Russell, personal observation, 3 October 1984).

SIZE: Attains 112-centimeter disc widths.

REFERENCES: Bearden 1961, 1965a; Bigelow and Schroeder 1953, 1962; Blaylock 1989; Gudger 1912; Joseph 1961; Radcliffe 1916; Schwartz 1959, 1989b, 1990, 1995, 2000a; Schwartz, Hogarth, and Weinstein 1982; Smith 1907; Smith and Merriner 1987; Wilson 1900.

Spotted eagle ray

Aetobatus narinari (Euphrasen, 1790)

DISTINGUISHING FEATURES: Eyes and spiracles on sides of head. Body nearly diamond shaped; edges of wings point posteriorly slightly behind midline across body. Head projects as a simple lobe. Tail whiplike with up to 8 serrated spines. Dorsal fin present. Pavement teeth in single series wide in both jaws, upper teeth flat; lower jaw V-shaped. Disc 2.1 times broad as long.

COLOR: Dark blue–black dorsally with white spots scattered over disk. White ventrally.

DISTRIBUTION AND OCCURRENCE: North Carolina to southern Brazil, Gulf of Mexico. Occurs in the Carolinas July–September. Will enter estuaries, sounds, and rivers.

SIZE: Attains 230-centimeter disc width.

REFERENCES: Bigelow and Schroeder 1953, 1962; Burton 1936; Gudger 1910, 1912; Jordan and Gilbert 1882; Radcliffe 1916; Schwartz 1984, 1995, 2000; Schwartz, Hogarth, and Weinstein 1982; Smith 1907; Yarrow 1877.

Bullnose ray

Myliobatis freminvillii Lesueur, 1824

DISTINGUISHING FEATURES: Eyes and spiracles on sides of head. Single broadly pointed lobe projects from head. Disc 1.7 times broad as long. Body naked. Whiplike tail with dorsal fin and serrated spine(s) near dorsal fin. Teeth white. Pavement teeth hexagonal in 7 series across in each jaw.

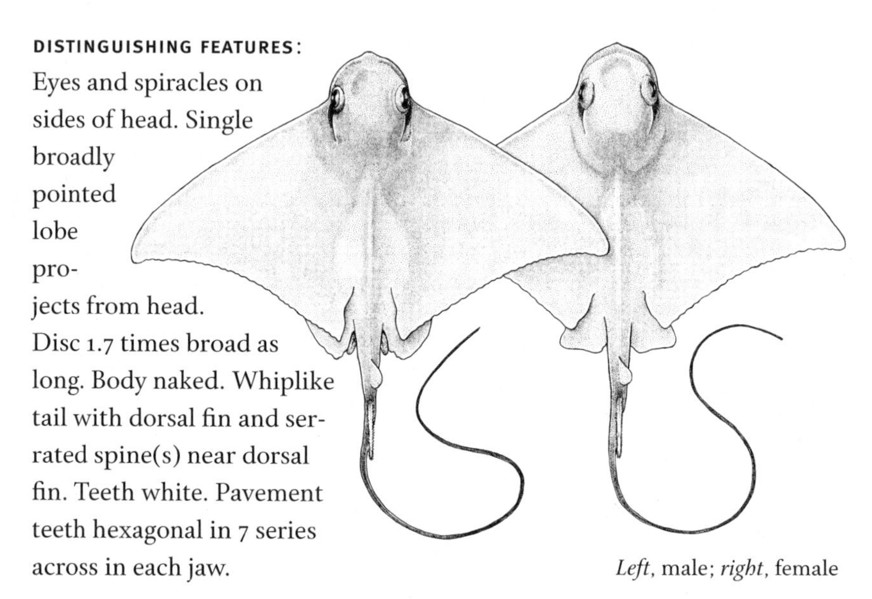

Left, male; *right*, female

COLOR: Chocolate brown dorsally; may have white or gray spots dorsally in fall or when stressed on capture. White ventrally.

DISTRIBUTION AND OCCURRENCE: Cape Cod, Massachusetts, to Brazil, Gulf of Mexico. Common in Carolinas May–October. Migrates seasonally north and south as waters warm and cool. Enters sounds.

SIZE: Attains 865-millimeter disc widths.

REFERENCES: Bearden 1961, 1965a; Bigelow and Schroeder 1953, 1962; Jenkins 1887; Radcliffe 1916; Schwartz 1984, 1995, 2000a; Smith 1907.

Southern eagle ray

Myliobatis goodei Garman, 1885

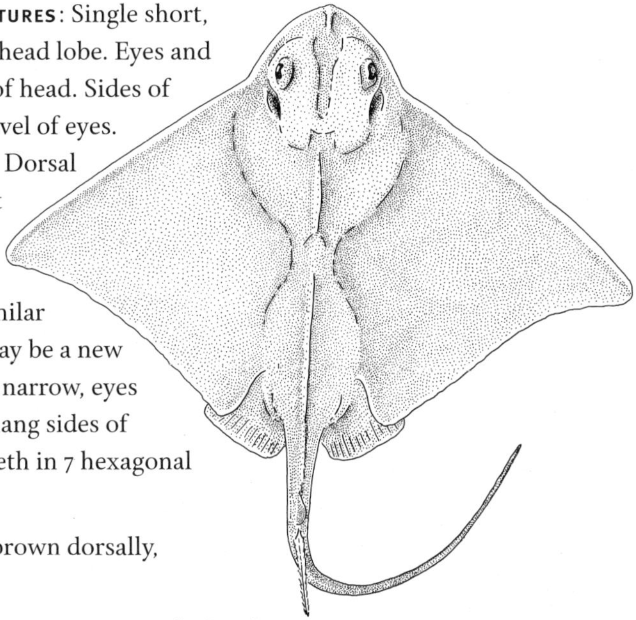

DISTINGUISHING FEATURES: Single short, protruding, broad head lobe. Eyes and spiracles on sides of head. Sides of head fall outside level of eyes. Dorsal fin present. Dorsal fin and spine(s) set away from rear of pelvic fin on whiplike tail. A similar form in the area may be a new species; its head is narrow, eyes level with or overhang sides of head. Pavement teeth in 7 hexagonal series.

COLOR: Chocolate brown dorsally, white ventrally.

DISTRIBUTION AND OCCURRENCE: South Carolina south to Argentina, South America in 54-meter waters.

SIZE: Attains 990-millimeter disc widths. Occurs rarely from South Carolina southward.

REFERENCES: Bearden 1965a; Bigelow and Schroeder 1953, 1962; Menni and Stehmann 2000; Schwartz 1984.

Nurse shark, *Ginglymostoma cirratum*
Doug Perrine/Seapics.com

Great hammerhead, *Sphyrna mokarran*
James D. Watt/Seapics.com

Great white shark, *Carcharodon carcharias*
James D. Watt/Seapics.com

Shortfin mako, *Isurus oxyrinchus*
Richard Herrmann/Seapics.com

Whale shark, *Rhincodon typus*
James D. Watt/Seapics.com

Blacknose shark, *Carcharhinus acronotus*
Doug Perrine/Seapics.com

Silky shark, *Carcharhinus falciformis*
Doug Perrine/Seapics.com

Bull shark, *Carcharhinus leucas*
Doug Perrine/Seapics.com

Blacktip shark, *Carcharhinus limbatus*
Doug Perrine/Seapics.com

Dusky shark, *Carcharhinus obscurus*
Rudie Kuiter/Seapics.com

Tiger shark, *Galeocerdo cuvier*
David B. Fleethan/Seapics.com

Lemon shark (3–4 years old), *Negaprion brevirostris*
Doug Perrine/Seapics.com

Spiny dogfish, *Squalus acanthias*
Scott Michael/Seapics.com

Atlantic guitarfish, *Rhinobatus lentiginosus*
Helmut Debelius/Seapics.com

Spotted eagle ray, *Aetobatis narinari*
Doug Perrine/Seapics.com

Pelagic stingray, *Pteroplatytrygon violacea*
Phillip Colla/Seapics.com

dichotomous keys

A dichotomous key is a series of couplets by which one selects between two alternatives. Eventually identification is achieved by following through the key in a step-by-step manner. For example, a tiger shark can be determined by following the choices in couplets 1a, 2b, 3b, 5b, 7b, 20b, 21b, 23b, 29b, 33b, 39b, and 41a. Thus, if a specimen meets the criteria in item 1a, proceed to couplet 2. If 2b is a match, proceed to couplet 3, and so forth.

Tooth counts or shapes (specific for each species) have not been stressed in the keys in order to determine identity of a species, as one rarely has time to examine and count the teeth of a thrashing, snapping shark. Colors likewise can vary by season and shark size and are not used in the key; however, see each species for color or size information. Do not place too much value on the presence of black-tipped pectoral fins, as several species possess these features. Additional information can be found among the species analyses, as similar features are usually found together in species descriptions, distributions, occurrences, sizes, and references.

Keys to Sharks, Skates, and Rays

1a. Gill slits located on lateral sides of head; anterior margin of pectoral fin not attached to side of head — 2

1b. Gill slits located ventrally on head; outer margin of pectoral fins attached to side of head — 57

2a. Body flattened dorso-ventrally; mouth terminal — *Squatina dumeril*, p. 33

2b. Body round in cross section; mouth ventral — 3

3a. Six or seven gill slits — 4

3b. Five gill slits — 5

4a. Six gill openings — *Hexanchus griseus*, p. 34

4b. Seven gill openings — *Heptranchias perlo*, p. 35

5a. Barbel present on nostril 6
5b. No barbel on nostril 7

6a. Barbels rudimentary or small, caudal peduncle keels, ridges along sides of body *Rhincodon typus* p. 57
6b. Barbels prominent, no caudal peduncle keels, no ridges along side of body *Ginglymostoma cirratum* p. 35

7a. Anal fin absent 8
7b. Anal fin present 20

8a. Clusters of denticles scattered over body *Echinorhinus brucus*, p. 36
8b. No clusters of denticles scattered over body 9

9a. No spines in dorsal fins 10
9b. Spines in dorsal fins 11

10a. Head length less than 20 percent of total length; body less than or equal to 160 centimeters long *Dalatias licha*, p. 38
10b. Head length greater than 20 percent of total length; body more than 160 centimeters long *Somniosus microcephalus*, p. 37

11a. Subcaudal keel on underside of caudal peduncle *Deania profundorum*, p. 38
11b. No subcaudal keel on underside of caudal peduncle 12

12a. Inner margin of pectoral fin elongate and pointed *Centrophorus granulosus*, p. 39
12b. Inner margin of pectoral fin not elongate and pointed 13

13a. No caudal fin keels, no precaudal pits 14
13b. Caudal fin keels, precaudal pits 17

14a. Interdorsal length equal to distance from snout tip to pectoral fin origin *Etmopterus hillianus*, p. 42
14b. Interdorsal length less than distance from snout tip to pectoral fin origin 15

15a. No black markings on body or tail *Centroscyllium fabricii*, p. 40
15b. Black markings on body and/or tail 16

16a. Caudal fin length greater than head length *Etmopterus bullisi*, p. 40
16b. Caudal fin length shorter or equal to head length *Etmopterus gracilispinis*, p. 41

17a. First dorsal fin and second dorsal fin equal in size; upper precaudal pit small or absent *Cirrhigaleus asper*, p. 42
17b. Second dorsal fin small, upper precaudal pit strong 18

18a. Anterior nasal flap single — *Squalus acanthias*, p. 43
18b. Anterior nasal flap double — 19

19a. Distance from fifth gill slit to first dorsal fin spine equal to distance from eye to second gill slit — *Squalus mitsukurii*, p. 45
19b. Distance from fifth gill slit to first dorsal fin spine less than distance from eye to second gill slit — *Squalus cubensis*, p. 44

20a. Gill slits long, extending almost full height of head — *Cetorhinus maximus*, p. 53
20b. Gill slits short, not extending full height of head — 21

21a. Caudal fin almost as long as rest of shark — 22
21b. Caudal fin shorter than rest of shark — 23

22a. Ridge along side of head giving appearance of helmet, eye higher than long — *Alopias superciliosus*, p. 51
22b. No ridge along side of head giving appearance of helmet, eye round — *Alopias vulpinus*, p. 52

23a. Caudal fin without lateral keels or with weak ones (not *Galeocerdo*) — 24
23b. Caudal fin with lateral keels — 29

24a. First dorsal fin posterior to midpoint of body — *Carcharias taurus*, p. 50
24b. First dorsal fin anterior to midpoint of body — 25

25a. No caudal peduncle keels — *Odontaspis ferox*, p. 51
25b. Caudal peduncle keels — 26

26a. Two caudal peduncle keels — *Lamna nasus*, p. 57
26b. One caudal peduncle keel — 27

27a. Origin of anal fin well behind ventral line of base of second dorsal fin; black spot usually present in axils of pectoral fins — *Carcharodon carcharias*, p. 54
27b. Origin of anal fin not ahead of second dorsal fin; no black spot in axils of pectoral fins — 28

28a. Pectoral fin length less than head length — *Isurus oxyrinchus*, p. 55
28b. Pectoral fin length greater than head length — *Isurus paucus*, p. 56

29a. Head flattened dorso-ventrally and expanded laterally — 30
29b. Head not flattened dorso-ventrally or expanded laterally — 33

30a. Head shovel-shaped — *Sphyrna tiburo*, p. 49
30b. Head hammer-shaped — 31

31a. Front margin of head notched at middle 32
31b. Front margins of head lacking notch at midline *Sphyrna zygaena*, p. 48

32a. Free posterior tip of second dorsal fin longer than vertical height of fin; mouth symphysis in advance of rear edge of head *Sphyrna lewini*, p. 46
32b. Free posterior tip of second dorsal fin shorter than fin height; mouth symphysis even with rear margin of head *Sphyrna mokarran*, p. 47

33a. Origin of first dorsal fin lies posterior to origin of pelvic fin 34
33b. Origin of first dorsal fin above or slightly anterior of axil of pectoral fin 39

34a. Body with chainlike or reticulated markings *Scyliorhinus retifer*, p. 62
34b. Body lacks chainlike markings 35

35a. Snout even or shorter than mouth width 36
35b. Snout longer than mouth width 38

36a. Body has white spots on body banding *Scyliorhinus hesperius*, p. 60
36b. Body has bars or blotches but not spots 37

37a. Mouth black *Galeus arae*, p. 60
37b. Mouth not black, body has dusky bars *Scyliorhinus meadi*, p. 61

38a. Denticles on upper surface of dorsal lobe of caudal fin, densely packed, forming crest *Apristurus profundorum*, p. 59
38b. Denticles on upper surface of dorsal lobe of caudal fin, identical, not forming crest *Apristurus laurussoni*, p. 58

39a. Precaudal pits absent 40
39b. Precaudal pits present 41

40a. Margin of lower lobe of caudal fin points posteriorly *Mustelus norrisi*, p. 64
40b. Margin of lower lobe of caudal fin not pointed posteriorly *Mustelus canis*, p. 63

41a. Lateral keels on caudal peduncle *Galeocerdo cuvier*, p. 70
41b. Lateral keels absent or weak on caudal peduncle 42

42a. Second dorsal fin nearly equal in size to first dorsal fin *Negaprion brevirostris*, p. 75
42b. Second dorsal fin smaller than first dorsal fin 43

43a. Second dorsal fin origin well behind anal fin origin *Rhizoprionodon terraenovae*, p. 77
43b. Second dorsal fin origin near anal fin origin 44

44a. Lateral keel on caudal peduncle *Prionace glauca*, p. 76
44b. No lateral keel on caudal peduncle 45

45a. Interdorsal ridge present between dorsal fins 46
45b. Interdorsal ridge absent between dorsal fins 53

46a. First dorsal fin high, broad, rounded, tipped white
Carcharhinus longimanus, p. 67
46b. First dorsal fin usually pointed, not rounded and not tipped white 47

47a. Snout long, eye bright green *Carcharhinus signatus*, p. 65
47b. Snout short, eye not bright green 48

48a. First dorsal fin origin well behind free rear tip of pectoral fin
Carcharhinus falciformis, p. 66
48b. First dorsal fin origin anterior to pectoral fin free near tip 49

49a. Origin of first dorsal fin just behind rear free tip of pectoral fin
Carcharhinus perezi, p. 70
49b. Origin of first dorsal fin over inner margin of pectoral fin 50

50a. First dorsal fin origin in front of pectoral fin insertion 51
50b. First dorsal fin opposite or near pectoral fin rear tip 52

51a. Distance from nostrils to mouth more than 2.4 times mouth width
Carcharhinus plumbeus, p. 69
51b. Distance from nostrils to mouth less than 2.4 times mouth width
Carcharhinus altimus, p. 64

52a. Second dorsal fin height less than 2.2 percent of total length
Carcharhinus obscurus, p. 68
52b. Second dorsal fin lower, more than 2.6 percent of total length
Carcharhinus galapagensis, p. 66

53a. Snout broad, rounded, first dorsal fin triangular, set over axil of pectoral fin
Carcharhinus leucas, p. 74
53b. Snout longer, more pointed, first dorsal fin located almost mid-body 54

54a. Black spot or smudge at tip of snout *Carcharhinus acronotus*, p. 71
54b. No black spot or smudge at tip of snout 55

55a. Snout long, V-shaped (viewed dorsally), first dorsal fin height more than 2.6 times interdorsal fin distance *Carcharhinus brevipinna*, p. 73
55b. Snout U-shaped (viewed dorsally), first dorsal fin height less than 2.6 times interdorsal fin distance 56

56a. No black tips to fins *Carcharhinus isodon*, p. 72
56b. Fins black-tipped *Carcharhinus limbatus*, p. 74

57a. Snout elongated as flat rostrum, edges armed with single row of teeth, lower lobe of caudal fin absent *Pristis pectinata*, p. 81
57b. Snout not flat rostrum 58

58a. Snout elongated and V-shaped (viewed dorsally), possesses distinct tail *Rhinobatos lentiginosus*, p. 82
58b. Snout not distinct V-shape (viewed dorsally) 59

59a. Possesses electric organs dorso-lateral and posterior to eyes 60
59b. Lacks electric organs dorso-lateral and posterior to eyes 61

60a. Mouth small, not surrounded by groove; front of disc straight or broadly rounded *Torpedo nobiliana*, p. 85
60b. Mouth small, surrounded by groove; disc rounded anteriorly 62

61a. Dorsal surface disc with black circles, blotches, or markings; found in shallow waters *Narcine brasiliensis*, p. 84
61b. Dorsal surface disc lacks markings, body very flaccid; found in deep water *Benthobatis marcida*, p. 83

62a. Two dorsal fins; tail stout, not armed with serrated spines, may have thorns or prickles 63
62b. One or no dorsal fin; tail well developed or whiplike, with or without serrated spines 77

63a. Outer margins of disc elongate forming forward-projecting lobes *Dactylobatus armatus*, p. 86
63b. Outer margin of disc pointed, no elongate forming forward-pointing lobes 64

64a. Mucous pores on lower surface of disc marked by dark dots or lines 65
64b. Mucous pores on lower surface of disc, not marked by dark dots or lines 66

65a. No interdorsal fin space; disc width 73 to 81 percent of total length; snout-mouth distance 19 to 25 percent of total length *Dipturus teevani*, p. 87
65b. Interdorsal fin space present; disc width 68 to 71 percent of total length; snout-mouth distance 17 to 18 percent of total length *Dipturus laevis*, p. 88

66a. Color brown and/or gray dorsally and ventrally 67
66b. Color brown dorsally, white, yellow, or dusky-black ventrally 68

67a. Distance between nostrils less than 10 percent of total length *Rajella bathyphila*, p. 89
67b. Distance between nostrils more than 10 percent of total length *Bathyraja richardsoni*, p. 90

68a. Dorsal fins confluent or separated by interdorsal spine 69
68b. Dorsal fins not separated by interdorsal spine, well separated with many small spines, dorsal fins black *Fenestraja atripinna*, p. 97

69a. Bars on tail *Fenestraja plutonia*, p. 91
69b. No bars on tail 70

70a. Three rows of conspicuous thorns along midportion of disc 71
70b. One or no rows of conspicuous thorns along midportion of disc 72

71a. Snout disc anterior margin angle 145° or more *Breviraja spinosa*, p. 92
71b. Snout disc anterior margin angle less than 145° *Malacoraja senta*, p. 95

72a. Snout disc angle about 120° *Neoraja carolinensis*, p. 96
72b. Snout disc angle less than 120° 73

73a. Snout long and clear; dorsal surface of disc marked with transverse or oblique bars *Raja eglanteria*, p. 100
73b. Snout blunt, not or very slightly clear; dorsal body surface with or without spots 74

74a. Dorsal surface of disc with rosette-like spots *Leucoraja garmani*, p. 99
74b. Dorsal surface of disc with or without spots, but not rosette-like 75

75a. Row of large thorns along tail, none larger than rest *Amblyraja radiata*, p. 98
75b. Three rows of large thorns along tail, none larger than rest 76

76a. Distance between eyes less than 12 percent of tail length *Leucoraja ocellata*, p. 93
76b. Distance between eyes more than 12 percent of tail length *Leucoraja erinacea*, p. 94

77a. Caudal fin well developed; tail with serrated spine(s); disc round *Urobatis jamaicensis*, p. 101
77b. Caudal fin present; tail whiplike with or without dorsal fin; tail with or without serrated spine(s) 78

78a. Eyes and spiracles on top of head 79
78b. Eyes and spiracles on sides of head 85

79a. Disc less than 1.5 times as broad as long 80
79b. Disc more than 1.5 times as broad as long 84

80a. Disc black or purple dorsally and ventrally; greatest disc width at level of eyes *Pteroplatytrygon violacea*, p. 102
80b. Disc white ventrally, greatest disc width well behind level of eyes 81

81a. Outer edges of disc rounded; maximum perpendicular disc width passes close to eyes 82
81b. Outer edges of disc not rounded; maximum perpendicular disc width passes over or behind eyes 83

82a. Snout anterior to eyes, longer than distance between spiracles *Dasyatis sabina*, p. 103
82b. Snout anterior to eyes, shorter than distance between spiracles *Dasyatis say*, p. 104

83a. Tail with low cutaneous fold on upper surface; small thorns on tail *Dasyatis americana*, p. 105
83b. Tail without low cutaneous fold on upper surface; large thorns or tubercles on tail *Dasyatis centroura*, p. 106

84a. No dorsal fin; tail with serrated spine(s); disc wider than long *Gymnura altavela*, p. 107
84b. No dorsal fin; tail without serrated spine(s); disc wider than long *Gymnura micrura*, p. 108

85a. Anterior edges of disc extend forward of head as separate cephalic fins 86
85b. Anterior edges of disc do not extend forward as separate cephalic fins 88

86a. Cephalic fins ahead of head, mouth terminal *Manta birostris*, p. 109
86b. Cephalic fins ahead of head, mouth on lower surface of head, not terminal 87

87a. Tail with spine; size greater than meter wide *Mobula mobular*, p. 111
87b. Tail without spine(s); size about a meter wide *Mobula hypostoma*, p. 110

88a. Teeth in each jaw as wide as jaw; body spotted dorsally; snout projects as one lobe; serrated spine(s) on tail *Aetobatus narinari*, p. 113
88b. Teeth in each jaw series 6 to 12; body unspotted; serrated spine(s) on tail 89

89a. Head bilobed with ventral fleshly flaps; tail with serrated spine(s) *Rhinoptera bonasus*, p. 112

89b. Head single lobed, pointed, without ventral fleshly flaps; tail with serrated spine(s) 90

90a. Dorsal fin near rear of pectoral fin; tail with serrated spine(s) *Myliobatis freminvillii*, p. 114

90b. Dorsal fin far back on whiplike tail, not near rear of pectoral fin; possess serrated spine(s) on tail *Myliobatis goodei*, p. 115

glossary

Albino: an organism that is all white and has pink eyes

Anal fin: unpaired fin located ventrally between the pelvic fins and the caudal fin

Axil: posterior end of fin base

Barbel: any soft extension hanging from lips or nostrils

Cartilaginous: a form of gristle not bone

Caudal fin crest: an arrangement of denticles along the upper surface of the caudal fin lobe

Caudal keel: a firm or fleshly expansion on the lateral aspect of the caudal peduncle or tail

Caudal peduncle: the body area between end of anal fin and base of caudal fin

Centrum: a bony segment of the backbone

Cephalic fin: finlike structure on either side of head (ventrally) formed by anterior portion of pectoral fins and lateral aspect of rostral cartilage

Cloaca: ventral body opening that contains the anal and the urogenital openings

Coastal: ocean area near land and on the continental shelf

Confluent: joined together

Continental shelf: ocean area from land seaward to where there is a steep drop into deep water

Cretaceous: a geological epoch occurring 66 to 98 million years ago

Cutaneous: pertaining to the skin

Denticles: an elasmobranch scale formed of an inner pulp cavity, middle dentine layer, and outer enamel layer; once formed does not grow further; usually ridged

Diplospondylous: two vertebrae in each body segment, centra smaller than monospondylous centra

Disc: a batoid condition where the body disc is flattened and completely joined to the head and pectoral fins

Dorsal fin anterior margin: front leading edge of dorsal fin

Dorsal fin height: vertical distance from dorsal fin base perpendicular to the body to the top of the fin

Dorsal fin posterior margin: rear edge of dorsal fin from top to rear extension near body

Electric organ: a honeycomb group of cells that produce electrical charges when activated

Estuaries: area within a river where saline and freshwater mix

Falcate: sickle or crescent shape to fin

Filter feeding: a state where gill filaments, because of their fine construction and attachment to the gill arches, filter out plankton

Fin spine: a stiff, sharp, pointed structure found in and supporting fins

Gill opening: one of five to seven openings on the side or ventral surface of the head

Head: length from tip of snout to last gill slit

Insertion: point of attachment of a fin to the body

Inshore: area of the ocean along the beach; depths are not great

Interdorsal distance: distance from first dorsal fin rear base to anterior base of second dorsal fin

Interorbital space: distance between inner edges of eyes

Lateral keel: longitudinal ridge along lateral aspect of caudal peduncle

Mesenchymal: mid-germinal embryonic layer from which muscular, vascular, and connective tissues develop

Monophyletic: a group of species where all of the descendants have a common ancestor

Monospondylous: single vertebra per body segment, centra larger than diplospondylous centra

Nictitating membrane: movable membrane along anterior ventral corner of eye in some species, especially requiem carcharhinids

Offshore: ocean area well away from land but not beyond continental shelf break into deep water

Oophagy: condition where internally developing embryo is nourished by eating fertilized or unfertilized eggs

Origin: anterior end of fin base

Oviparous: external development of an embryo; fertilized egg development takes place in egg case outside the female

Ovoviviparous: internal development of embryo, and for the most part embryo is reliant on yolk to develop

Paleozoic: geologic era about 400 to 600 million years ago

Pelagic: open ocean area far from land

Pelvic fin anterior margin: leading edge of pelvic fin

Pelvic fin inner margin: part of fin next to body of shark

Pelvic fin insertion or base: place of attachment to body

Pelvic fin posterior margin: part of fin tip to rear point of fin near body

Placenta: connection between female and embryo in uterus that permits nutrient and waste transfer

Precaudal pit: transverse groove at front base of caudal fin upper or lower lobe

Rosette: cluster of spots arranged in a circle

Rostrum: a projecting snout, projecting from anterior part of head, as in sawfish

Sand shark: local Carolinian name for small, gray-colored sharks, regardless of species

Serrate: sawlike notches on tail spines of stingrays

Snout: area of head ahead of nostrils

Sound: large body of water lying between the mainland and a barrier island

Spine(s): sharp, stiff, unsegmented projection(s) found at front of dorsal rays or serrated on tail of stingrays

Spiracle: dorsal surface opening behind eye for water entry into gill chamber during breathing

Terminal mouth: mouth that is located at the front of the head

Thorn: a small or large denticle that develops a very sharp point, usually found in skates and rays

Total length: maximum distance from tip of the snout to the very end of the tail (held in a natural position and not bent to extend the length)

Toxin: a poison produced by select cells found in ventral grooves of stingray spines

Viviparous: internal development where female provides nutrients to the embryo and gives birth to living young

Wing width: maximum distance across body from one pectoral wing edge to the other

references

Able, K. W., and D. Flescher. 1991. Distribution and habitat of the chain dogfish, *Scyliorhinus retifer*, in the mid-Atlantic bight. Copeia 1991(1):233–234.

Adams, W. T., and C. Wilson. 1995. The status of the smalltooth sawfish, *Pristis pectinata* Latham, 1794 (Pristiformes: Pristidae), in the United States. Chondros 6(4):1–5.

Al-Badri, M., and R. Lawson. 1985. Contribution to the taxonomy of the spiny dogfish, *Squalus acanthias* L. Cybium, 3rd ser., 9(4):385–399.

Anderson, E. D. 1990. Fishery methods as applied to elasmobranch fisheries. Pp. 473–484 in Elasmobranchs as living resources: advances in the biology, ecology, systematics, and the status of the fisheries. H. L. Pratt Jr., S. H. Gruber, and T. Taniuchi (eds.). NOAA Technical Report NMFS 90.

Anderson, R. C., and J. D. Stevens. 1996. Review of information on diurnal vertical migrations in the bignose shark (*Carcharhinus altimus*). Marine Freshwater Research 47(4):605–608.

Anonymous. 1998. Shark fisheries, management and biology. Marine Freshwater Research 49(7):553–767.

Backus, R. H., S. Springer, and L. Arnold. 1956. A contribution to the natural history of the white-tip shark, *Pterolamiops longimanus* (Poey). Deep Sea Research 3:178–188.

Barans, C. A., and G. F. Ulrich. 1996. Sixgill shark, *Hexanchus griseus*, aggressive feeding behavior on epibenthic crabs. Journal of the Elisha Mitchell Scientific Society 110(3–4):49–52.

Bass, A. J., J. D. D'Aubrey, and N. Kistanasamy. 1973. Sharks of the east coast of Southern Africa I. The genus *Carcharhinus* (Carcharhinidae). Investigation Report of the Oceanographic Research Institute 33:1–168.

———. 1976. Sharks of the east coast of Southern Africa VI. The families Oxynotidae, Squalidae, Dalatidae, and Echinorhinidae. Investigation Report of the Oceanographic Research Institute 45:1–103.

Bean, B. A., and A. C. Weed. 1909. Description of a new skate (*Dactylobatus armatus*) from deep water off the southern Atlantic coast of the United States. Proceedings of the United States National Museum 36:459–461.

Bearden, C. M. 1961. Common fishes of South Carolina. Contributions, Bears Bluff Laboratory, no. 34. 47 pp.
———. 1965a. Elasmobranch fishes of South Carolina. Contributions, Bears Bluff Laboratory, no. 42. 22 pp.
———. 1965b. Occurrence of the spiny dogfish, *Squalus acanthias*, and other elasmobranchs in South Carolina coastal waters. Copeia 1965(3):378.
Benz, F. W., and J. A. Caira. 1995. Eels from the heart of a mako shark. Association of Southeastern Biologists Bulletin 42(2):153.
Bigelow, H. B., and W. C. Schroeder. 1940. Sharks of the genus *Mustelus* in the western Atlantic. Proceedings of the Boston Society of Natural Sciences 41:417–438.
———. 1944. New sharks from the western North Atlantic. Proceedings of the New England Zoology Club 23:21–36.
———. 1948. Status of the western North Atlantic: lancelets, cyclostomes, and sharks. Sears Foundation Marine Research Memoir 1. 576 pp.
———. 1950. New and little-known cartilaginous fishes from the Atlantic. Bulletin of the Museum of Comparative Zoology, Harvard 103:385–406.
———. 1951. Three new skates and a new chimaeroid from the Gulf of Mexico. Journal of the Washington Academy of Sciences 41:383–392.
———. 1953. Fishes of the western North Atlantic: sawfishes, guitarfishes, skates, rays, and chimeras. Pt. 2. Sears Foundation Marine Research Memoir 1. 588 pp.
———. 1954. Deepwater elasmobranchs and chimaeroids from the northwestern Atlantic slope. Bulletin of the Museum of Comparative Zoology, Harvard 112:38–87.
———. 1957. A study of the sharks of the suborder Squaloidea. Bulletin of the Museum of Comparative Zoology, Harvard 117(1):1–150.
———. 1961. Life history notes on the roughtail stingray *Dasyatis centroura* (Mitchill). Copeia 1961:232–234.
———. 1962. New and little-known batoid fishes from the western Atlantic. Bulletin of the Museum of Comparative Zoology, Harvard 128:162–244.
———. 1965. A further account of the batoid fishes from the western Atlantic. Bulletin of the Museum of Comparative Zoology, Harvard 132(5):465–477.
———. 1968a. Additional notes on batoid fishes from the western north Atlantic. Breviora, no. 281. 23 pp.
———. 1968b. New records of two geographically restrictive species of western Atlantic skates, *Breviraja yucatanensis* and *Dactylobatus armatus*. Copeia 1968(3):630–631.
Bigelow, H. B., W. C. Schroeder, and S. Springer. 1953. New and little-known sharks from the Atlantic and from the Gulf of Mexico. Bulletin of the Museum of Comparative Zoology, Harvard 109:213–276.
———. 1955. Three new shark records from the Gulf of Mexico. Breviora 49:1–12.

Blackman, R. R. 1972. Occurrence of a new skate in the western north Atlantic. Quarterly Journal of the Florida Academy of Sciences 35(1):27–28.

Blaylock, R. A. 1989. A massive school of cownose rays, *Rhinoptera bonasus*, in the lower Chesapeake Bay, Virginia. Copeia 1989:744–748.

Bonfil, R. S. 1989. The abnormal embryo of the reef shark, *Carcharhinus perezi* (Poey), from Yucatan, Mexico. Northeast Gulf Science 10(2):153–155.

———. 1994. Review of world elasmobranch fisheries. FAO Fisheries Technical Paper, no. 341. Rome. 119 pp.

Branstetter, S. 1982. Problems associated with the identification and separation of the spinner shark, *Carcharhinus brevipinna*, and the blacktip shark, *Carcharhinus limbatus*. Copeia 1982(4):421–464.

Branstetter, S., and J. D. McEachran. 1983. A first record of the bigeye thresher, *Alopias superciliosus*, the blue shark, *Prionace glauca*, and the pelagic stingray, *Dasyatis violacea*, from the Gulf of Mexico. Northeast Gulf Science 6:59–61.

Brimley, H. H. 1935a. Basking sharks (*Cetorhinus maximus*) in North Carolina waters. Journal of the Elisha Mitchell Scientific Society 51:311.

———. 1935b. Notes on the occurrence of a whale shark (*Rhincodon typus*) in the Cape Fear river near Southport, North Carolina. Journal of the Elisha Mitchell Scientific Society 51:160–162.

Bullis, H. R., Jr. 1967. Depth segregation and distribution of sex mature groups in the marbled catshark, *Galeus arae*. Pp. 141–174 in Sharks, skates, and rays. P. W. Gilbert, R. F. Mathewson, and D. P. Rall (eds.). Johns Hopkins University Press, Baltimore, Md.

Bullis, H. R., Jr., and P. Struhsaker. 1961. Life history notes on the roughtail stingray, *Dasyatis centroura* (Mitchill). Copeia 1961:232–234.

Burgess, G. H., G. W. Link Jr., and S. W. Ross. 1979. Additional marine fishes new or rare to Carolina waters. Northeast Gulf Science 3(2):74–87.

Burton, E. M. 1936. Marine fishes of South Carolina. South Carolina vertebrate fauna. N. Gist (ed.). Gel Lander Collection.

Campbell, R. A., and T. A. Monroe. 1974. Discovery of the lesser devil ray, *Mobula hypostoma*, in southern New England. Chesapeake Science 15(2):114–115.

Carey, F. G., and E. Clark. 1995. Depth telemetry from the sixgill shark, *Hexanchus griseus*, at Bermuda. Environmental Biology of Fishes 42(1):7–15.

Carter, J. G., P. E. Gallagher, R. E. Valone, T. J. Rossback, P. G. Gensol, W. H. Wheeler, and D. Whitman. 1988. Fossil collecting in North Carolina. Department of Natural Resources Commission and Development, Division of Land Resources, Geological Survey Section, Raleigh, N.C. Bulletin 89. 89 pp.

Case, G. R. 1982. A pictorial guide to fossils. Van Nostrand, New York. 514 pp.

Casey, J., and R. A. Myers. 1998. Near extinction of a large widely distributed fish. Science 281:690–692.

Casey, J. G., and H. L. Pratt, Jr. 1985. Distribution of the white shark, *Carcharodon carcharias*, in the western north Atlantic waters. Memoirs of the Southern California Academy of Sciences 9:2–14.

Casey, J. G., F. J. Mather III, and J. Hoenig. 1978. Offshore fisheries of the middle Atlantic bight. Marine Freshwater Research 3:107–129.

Castro, J. I. 1993a. The biology of the finetooth shark, *Carcharhinus isodon*. Environmental Biology of Fishes 36:219–232.

———. 1993b. The shark nursery of Bulls Bay, South Carolina, with a review of the shark nurseries of the southeastern coast of the United States. Environmental Biology of Fishes 38(1–3):37–48.

———. 1996a. Biology of the blacktip shark, *Carcharhinus limbatus*, off the southeastern United States. Bulletin of Marine Science 59(3):508–522.

———. 1996b. The sharks of North American waters. 2nd ed. Texas A&M University Press, College Station. 180 pp.

———. 2000. The biology of the nurse shark, *Ginglymostoma cirratum*, off the Florida east coast and the Bahama Islands. Environmental Biology of Fishes 58:1–32.

Castro, J. I., C. M. Woodley, and R. L. Brudek. 1999. A preliminary evaluation of the status of shark species. FAO Fisheries Technical Paper, no. 380. Rome. 72 pp.

Choi, Y., I.-S. Kim, and K. Nakaya. 1998. A taxonomic revision of the *Carcharhinus* (Pisces: Elasmobranchi), with a description of two new records in Korea. Korean Journal of Systematic Zoology 14(1):43–49.

Clark, E., and E. Kristoff. 1990. Deep-sea elasmobranchs observed from submersibles off Bermuda, Grand Cayman, and Freeport, Bahamas. Pp. 169–184 in Elasmobranchs as living resources: advances in the biology, ecology, systematics, and the status of the fisheries. H. L. Pratt Jr., S. H. Gruber, and T. Taniuchi (eds.). NOAA Technical Report NMFS 90.

Clark, T. A. 1984. The ecology of the scalloped hammerhead, *Sphyrna lewini*, in Hawaii. Pacific Science 25:133–144.

Coles, R. J. 1910. Observations on the habits and distribution of certain fishes taken on the coast of North Carolina. Bulletin of the American Museum of Natural History 28:337–348.

———. 1913. Notes on the embryos of several species of rays, with remarks on the northward summer migration of certain tropical tunas observed on the coast of North Carolina. Bulletin of the American Museum of Natural History 32(2):29–35.

———. 1915. Notes on the sharks and rays of Cape Lookout, North Carolina. Proceedings of the Biological Society of Washington 28:89–94.

———. 1916. Natural history notes on the devil fish, *Manta birostris* (Walbaum), and *Mobula olfersi* (Müller). Bulletin of the American Museum of Natural History 35:648–657.

———. 1919. The large sharks of Cape Lookout, North Carolina, the white shark or maneater, tiger shark, and hammerhead. Copeia 69:34–43.

———. 1926. Notes on Cape Lookout (North Carolina) fishes—1925. Copeia 151:105–106.

Compagno, L. J. V. 1973. Interrelationships of living elasmobranchs. Zoological Journal of the Linnaean Society of London 53(suppl. 1):315–361.

———. 1977. Phyletic relationships of living elasmobranchs. American Zoologist 17(2):303–322.

———. 1984. Sharks of the world. FAO Species Catalogue, Species Synopsis 125, vol. 4, pt. 1:1–250; pt. 2:251–655. Rome.

———. 1988. Sharks of the order Carcharhiniformes. Princeton University Press, Princeton, N.J. 486 pp.

———. 1990. Alternative life-history styles of cartilaginous fishes in time and space. Environmental Biology of Fishes 28:33–75.

———. 1991. The evolution and diversity of sharks. Pp. 15–22 in Discovering sharks. S. H. Gruber (ed.). American Littoral Society, Highlands, N.J.

———. 1999. Systematics and body form. Pp. 1–42 in Sharks, skates, and rays: the biology of elasmobranch fishes. W. C. Hamlett (ed.). Johns Hopkins University Press, Baltimore, Md.

———. 2001. Sharks of the world, an annotated and illustrated catalogue of shark species known to date: bullhead, mackerel, and carpet sharks (Heterodontiformes, Lamniformes, and Orectolobiformes). Food and aquacultural organization. Species Catalogue Fishery Purposes, vol. 1, no. 2. Rome. 269 pp.

Compagno, L. J. V., and S. F. Cook. 1995. The exploitation and conservation of fresh water elasmobranchs: status of taxa and prospects for the future. Journal of Aquariculture and Aquatic Sciences 7:62–90.

Compagno, L. J. V., G. D. Zorzi, H. Ishihara, and J. Caira. 1990. Recommendations for future research in systematics, geographic distribution, and evolutionary biology of Chondrichthyan fishes. Pp. 513–515 in Elasmobranchs as living resources: advances in the biology, ecology, systematics, and the status of the fisheries. H. L. Pratt Jr., S. H. Gruber, and T. Taniuchi (eds.). NOAA Technical Report NMFS 90.

Daiber, F. C. 1959. On observations on the deep-sea electric ray. Copeia 1959(1):74.

de Carvalho, M. R. 1996. Higher-level elasmobranch phylogenies, basal squaleans, and paraphyly. Pp. 35–62 in Interrelationships of fishes. M. L. J. Stiassny, L. R. Parenti, and G. D. Johnson (eds.). Academic Press, San Diego.

———. 1999. A synopsis of the deep-sea genus *Benthobatis* Alcock, with a redescription of the type species *Benthobatis moresbyi* Alcock, 1898 (Chondrichthyes, Torpediniformes, Narcinidae). Pp. 231–255 in Proceedings of the Fifth Indo-Pacific Fish Conference, Nouméa, New Caledonia, 3–8 November 1997. B. Seret and J.-Y. Sire (eds.). Société Franáaise d'Ichthyologie, Paris.

Dingerkus, G. 1986. Interrelationships of orectolobiform sharks (Chondrichthyes: Selachii). Pp. 227–245 in Indo-Pacific fish biology: Proceedings of the Second International Conference on Indo-Pacific Fishes. T. Uyeno, R. Arai, T. Taniuchi, and K. Matsuura (eds.). Ichthyological Society of Japan, Tokyo.

———. 1995. Relationships of Potamotrygoniform stingrays. Journal of Aquariculture and Aquatic Sciences 7:32–37.

Dodril, J. W., and R. G. Gilmore. 1979. First North American record of the longfin mako, *Isurus paucus* (Guitart Manday). Florida Scientist 42:152–158.

Dulvy, N. K., and J. D. Reynolds. 2002. Predicting extinction vulnerability of skates. Conservation Biology 16(2):440–450.

Dulvy, N. K., J. D. Metcalfe, J. Glanville, M. G. Pawson, and J. D. Reynolds. 2000. Fishery stability, local extinctions, and shifts in community structure in skates. Conservation Biology 14(1):283–294.

Dunn, K. A., and J. F. Morissey. 1995. Molecular phylogeny of elasmobranchs. Copeia 1995(3):526–535.

Ellis, R., and J. E. McCosker. 1991. Great white shark. Harper Collins, New York. 270 pp.

Feduccia, A., and B. H. Slaughter. 1974. Sexual dimorphism in skates (Rajidae) and its possible role in differential niche utilization. Evolution 28(1):164–468.

Frazzetta, T. H. 1988. The mechanisms of cutting and the form of shark teeth (Chondrichthyes, Elasmobranchii). Zoomorphology 108:98–107.

Frøiland, O. 1975. Albinisme hos hai. Fauna, Oslo 28(3):170–173.

Gadiz, O. B. F., A. Medini, M. A. Bezera, and M. A. N. Furtendo Neto. 1999. Data on *Squatina dumerili* (Chondrichthyes, Squatinidae) from Brazil, with taxonomic comments on the genus *Squatina* of the Brazilian coast. Arquivos de Ciencias do Mar Forteleza 32:133–136.

Garman, S. 1913. The Plagiostomia (sharks, skates, and rays). Memoirs of the Museum of Comparative Zoology 36:1–528.

Garrick, J. A. F. 1960a. Studies on New Zealand elasmobranchii. Pt. 10, The genus *Echinorhinus*, with an account of a second species, *E. cookei* Pietschmann, 1928. Transactions of the Royal Society of New Zealand 88(1):105–117.

———. 1960b. Studies on New Zealand elasmobranchii. Pt. 11, Squaloids of the genera *Deania*, *Etmoptera*, *Oxynotus*, and *Dalatias* in New Zealand waters. Transactions of the Royal Society of New Zealand 88(3):489–517.

———. 1960c. Studies on New Zealand elasmobranchii. Pt. 12, The species of *Squalus* from New Zealand and Australia, and a general account and key to the New Zealand Squaloidea. Transactions of the Royal Society of New Zealand 88(3):519–557.

———. 1961. Studies on New Zealand elasmobranchii. Pt. 13, A new species of *Raja* from 1,300 fathoms. Transactions of the Royal Society of New Zealand 88(4):743–748.

———. 1967a. A broad view of *Carcharhinus* species, their systematics and distribution. Pp. 85–91 in Sharks, skates, and rays. P. W. Gilbert, R. F. Matthewson, and D. P. Rall (eds.). Johns Hopkins University Press, Baltimore, Md.

———. 1967b. Revision of the sharks of the genus *Isurus* with description of a new species (Galeoidea, Lamnidae). Proceedings of the United States National Museum 118(3537):663–690.

———. 1982. Sharks of the genus *Carcharhinus*. NOAA Technical Report NMFS Circular 445. 194 pp.

Garrick, J. A. F., and L. J. Paul. 1971. *Heptranchus dakin* Whitley, 1931, a synonym of *Hexanchus perlo* (Bonnaterre, 1788), the sharpnosed sevengill or perlon shark, with notes on sexual dimorphism in this species. Zoological Publication of Victoria University, no. 54. Wellington, New Zealand. 14 pp.

Garrick, J. A. F., and L. P. Schultz. 1963. A guide to the kinds of potentially dangerous sharks. Pp. 3–62 in Sharks and survival. P. W. Gilbert (ed.). D. C. Heath, Boston.

Garrick, J. A. F., R. H. Backus, and R. H. Gibbs Jr. 1964. *Carcharhinus floridanus*, the silky shark, a synonym of *C. falciformis*. Copeia 1964(2):369–375.

Gilbert, C. R. 1967a. A revision of the hammerhead sharks (family Sphyrnidae). Proceedings of the United States National Museum 119(3539):1–88.

———. 1967b. A taxonomic synopsis of the hammerhead sharks (family Sphyrnidae). Pp. 69–83 in Sharks, skates, and rays. P. W. Gilbert, R. F. Mathewson, and D. P. Rall (eds.). Johns Hopkins University Press, Baltimore, Md.

Gilbert, P. W. 1963. Sharks and survival. D. C. Heath, Boston. 578 pp.

Gilken, J., and B. W. Coad. 1989. The bluntnose sixgill shark, *Hexanchus griseus* (Bonnaterre, 1788), new to the fish fauna of Atlantic Canada. Proceedings of the Nova Scotia Institute of Sciences, Halifax 39(2):75–77.

Gill, T. 1903. The devilfish and some other fishes in North Carolina. Forest and Stream 9:431.

Goto, T. 2001. Comparative anatomy, phylogeny, and cladistic classification in the order Orectolobiformes (Chondrichthyes, Elasmobranchii). Memoirs of the Graduate School of Fisheries Sciences, Hokkaido University, Hakodate, Japan 48(1):1–100.

Gruber, S. H., and L. V. J. Compagno. 1983. Taxonomic studies and biology of the bigeye thresher, *Alopias superciliosus*. Fishery Bulletin 79:617–640.

Gudger, E. W. 1907. A note on the hammerhead shark (*Sphyrna zygaena*) and its food. Science 25:1005–1006.

———. 1910. Notes on some Beaufort, North Carolina, fisheries, 1909. American Naturalist 44:395–404.

———. 1912. Natural history notes on some Beaufort, North Carolina, fisheries, 1910–11. Proceedings of the Biological Society of Washington 25:141–156.

———. 1913a. Natural history notes on some Beaufort, North Carolina fishes,

1910–11. III. Fishes new or little known on the coast of North Carolina. Journal of the Elisha Mitchell Scientific Society 28(44):157–172.

———. 1913b. Natural history notes on some Beaufort, North Carolina fishes, 1912. Proceedings of the Biological Society of Washington 26:97–109.

———. 1915. Natural history of the whale shark, *Rhincodon typus* Smith. Zoologia, New York 1(19):349–389.

———. 1932. Cannibalism among the sharks and rays. Scientific Monthly 34:403–419.

———. 1943. Is the sting ray's spine poisonous? A historical resumé showing the development of our knowledge that it is poisonous. Bulletin of Historical Medicine 14:467–504.

———. 1947. Sizes attained by the larger hammerhead sharks. Copeia 1947(4):228–236.

———. 1948a. The basking shark *Cetorhinus maximus* on the North Carolina coast. Journal of the Elisha Mitchell Scientific Society 29(1):41–44.

———. 1948b. The tiger shark *Galeocerdo tigrinus* on the North Carolina coast. Journal of the Elisha Mitchell Scientific Society 64:221–223.

———. 1949. Natural history notes on tiger sharks, *Galeocerdo tigrinus* Smith. Copeia 1949:39–47.

Guitart Manday, D. 1966. Nuevo nombre para una especie de tiburon del genero *Isurus* (Elasmobranchii: Isuridae) de aquas cubanas. Poeyana (ser. A) 1966(15):1–9.

Hall, B. 1982. Bone in the cartilaginous fishes. Nature, London 298:324.

Halstead, B. W. 1970. Venomous marine animals of the world. Vol. 3, Vertebrates. United States Printing Office, Washington, D.C. 1,004 pp.

Hamlett, W. C. 1999. Sharks, skates, and rays: the biology of elasmobranch fishes. W. C. Hamlett (ed.). Johns Hopkins University Press, Baltimore, Md. 575 pp.

Hamlett, W. C., and T. J. Koob. 1999. Female reproductive system. Pp. 398–493 in Sharks, skates, and rays: the biology of elasmobranch fishes. W. C. Hamlett (ed.). Johns Hopkins University Press, Baltimore, Md.

Heemstra, P. C. 1997. A review of the smooth-hound sharks (Genus *Mustelus*; family Triakidae) of the western Atlantic Ocean, with description of two new species and a new subspecies. Bulletin of Marine Science 60(3):894–928.

Heemstra, P. C., and M. M. Smith. 1980. Hexatrygonidae, a new family of stingrays (Myliobatiformes: Batoidea) from South Africa, with comments on the classification of batoid fishes. J. L. B. Smith Institute Ichthyology Bulletin 43. 17 pp.

Herdendorf, C. E., and T. M. Berra. 1995. A Greenland shark from the wreck of the S.S. Central America at 2,200 meters. Transactions of the American Fisheries Society 124(6):950–953.

Hildebrand, S. F., and W. C. Schroeder. 1928. Fishes of Chesapeake Bay. Bulletin of the United States Bureau of Fisheries 43(1):1–366.

Hoenig, J. M., and S. H. Gruber. 1990. Life-history patterns in the elasmobranchs: implications for fisheries management. Pp. 1–16 in Elasmobranchs as living resources: advances in the biology, ecology, systematics, and the status of the fisheries. H. L. Pratt Jr., S. H. Gruber, and T. Taniuchi (eds.). NOAA Technical Report NMFS 90.

Hoenig, J. M., and A. H. Walsh. 1983. Skeletal lesions and deformities in large sharks. Journal of Wildlife Diseases 19(1):27–33.

Hoff, T. B., and J. A. Musick. 1990. Western North Atlantic sharks: fishery management problems and informational requirements. Pp. 455–472 in Elasmobranchs as living resources: advances in the biology, ecology, systematics, and the status of the fisheries. H. L. Pratt Jr., S. H. Gruber, and T. Taniuchi (eds.). NOAA Technical Report NMFS 90.

Holden, M. J. 1974. Problems in the rational exploitation of elasmobranch populations and suggested solutions. Pp. 117–137 in Sea Fisheries Research. F. R. Harden-Jones (ed.). Paul Dek, London.

Hueter, R. E. 1991. Survey of the Florida recreational shark fishery utilizing shark tournament and selective longline data. Final Report, Florida Department of Natural Resources Grant, FDNR Grant Agreement 6627. Mote Marine Laboratory, Sarasota, Fla. 74 pp.

———. 1998. Science and management of shark fisheries. Fisheries Research 39(2):1–228.

Huish, M. T., and C. Benedict. 1977. Some tracking of dusky shark in the Cape Fear River, North Carolina. Journal of the Elisha Mitchell Scientific Society 93(1):28–36.

International Game Fish Association. 1998. World record game fishes, 1998. International Game Fish Association, Pompano Beach, Fla. 352 pp.

Ishihara, H., K. Honma, and R. Nakamura. 2001. Albinism individuals of manta rays and Japanese common skates found in the western North Pacific. Report of the Japanese Society of Elasmobranch Studies, no. 37:36.

Ishihara, H., T. Taniuchi, and N. Shizmizu. 1991. Sexual description of numbers of rostral teeth in the sawfish, *Pristis microdon*, collected from Australia and Papua, New Guinea. University of Tokyo Museum. Nature Culture 3:83–89.

Jacob, B. A., and J. D. McEachran. 1994. Status of two species of skates *Raja (Dipturus) teevani* and *Raja (Dipturus) floridana* (Chondrichthyes: Rajoidei) from the western North Atlantic. Copeia 1994(2):433–445.

Jenkins, O. P. 1885. Notes on the fishes of Beaufort Harbor, North Carolina. Johns Hopkins University Circular (October 1885):11.

———. 1887. A list of fishes of Beaufort Harbor, North Carolina. Studies from the Biological Laboratory, Johns Hopkins University 4:88–94.

Jensen, A. C. 1966. Life history of the spiny dogfish, *Squalus acanthias*. Fishery Bulletin 65(3):537–554.

Jensen, C. F. 1998. Cooperative Atlantic states shark pupping and nursery survey (Coastspan) in North Carolina, May–November 1998. Report of the Apex Predatory Program. 21 pp.

Jensen, C. F., and G. A. Hopkins. 2001. Evaluation of bycatch in North Carolina Spanish and king mackerel sinknet fishery, with emphasis on sharks during October and November 1998 and 2000, including historical data from 1996–1997. Report of the Sea Grant Program 98 FEG-47. Raleigh, N.C. 63 pp.

Jensen, C. F., L. J. Natanson, H. L. Pratt, N. E. Kohler, and S. Compana. 2002. The reproductive biology of the porbeagle shark, *Lamna nasus*, in the western North Atlantic Ocean. Fishery Bulletin 100(4): 727–738.

Joel, J. Z., and L. P. Ebenzer. 1991. On the bramble shark, with 52 embryos. Indian Council Aquaculture Research, Marine Fisheries Information Services Technical Extension Series, no. 108:15, 31.

Jordan, D. S., and C. H. Gilbert. 1882. Notes on a collection of fishes from Charleston, South Carolina, with description of three new species. Proceedings of the United States National Museum 5:580–620.

Joseph, E. B. 1961. An albino cownose ray, *Rhinoptera bonasus* (Mitchill), from Chesapeake Bay. Copeia 1961(4):482–483.

Joung, S.-J., C.-T. Chen, E. Clark, S. Uchida, and W. Y. P. Huang. 1996. The whale shark, *Rhincodon typus*, is a livebearer: 300 embryos found in the "megamamma" supreme. Environmental Biology of Fishes 46(3):219–223.

Kajura, S. M., and T. C. Tricas. 1996. Seasonal dynamics of dental sexual dimorphism in the Atlantic stingray, *Dasyatis sabina*. Journal of Experimental Biology 119(10):2297–2306.

Kemp, N. E. 1999. Integumentary system and teeth. Pp. 43–68 in Sharks, skates, and rays: the biology of elasmobranch fishes. W. C. Hamlett (ed.). Johns Hopkins University Press, Baltimore, Md.

Kent, B. W. 1994. Fossil sharks of the Chesapeake Bay area. Egan Rees and Boyer, Columbia, Md. 146 pp.

———. 1999a. Pt. 2, Sharks from the Fisher/Sullivan site, in early Eocene vertebrates and plants from the Fisher/Sullivan site (Nanjemoy formation), Stafford County, Virginia. Pp. 11–37 in Virginia Division of Mineral Resources Publication 152.

———. 1999b. Pt. 3, Rays from the Fisher/Sullivan site. Pp. 39–51 in Virginia Division of Mineral Resources Publication 152.

Killam, K., and G. Parsons. 1986. First record of the longfin mako, *Isurus paucus*, in the Gulf of Mexico. Fishery Bulletin 84(3):748–799.

Kohler, N. E., J. G. Casey, and P. R. Turner. 1998. NMFS Cooperative Shark Tagging Program, 1962–1993. Marine Fisheries Review 60(2):1–87.

Krefft, G. 1968a. Knorpelfisch (Chondrichthyes) aus dem Trogischen ostatlantic. Atlantide Report 10:33–76.
———. 1968b. Neue und erstmalig nachgewiesene Knorpelfische aus dem archibenthal des Südwestatlantiks, einschliezlick einer Diskussion eineger *Etmopterus*-Arten Südlicher Meere. Archives Fischereiwissenschaften 19(1):1–42.
———. 1980. Result of the research cruises of FRV *Walter Herwig* to South America, 53 sharks from the pelagic trawl catch obtained during Atlantic transects, including some species from other cruises. Archives Fischereiwissenschaften 30(1):1–16.
Lawson, J. 1709. A new voyage to Carolina, containing the exact description and natural history of that country. London. 171 pp.
Litvinov, F. F. 1982. Two forms of teeth in the blue shark, *Prionace glaucus*, Carcharhinidae. Journal of Ichthyology 22(4):154–156.
———. 1983. Rate of teeth replacement in the blue shark, *Prionace glaucus* (Carcharhinidae), in relation to feeding. Voprosy Ikthyologia 23(1):143–145 (translation).
Liu, K.-M., C.-T. Chen, and S.-J. Joung. 2001. A study of shark resources in the waters off Taiwan. Pp. 249–256 in Aquaculture and fisheries resources management. I. C. Liao and J. Baker (eds.). Joint Taiwan-Australia Aquaculture Fisheries Resources Management Forum, TFRI Conference Proceedings 4, Taiwan Fisheries Research Institute, Keelung.
Lovejoy, N. R. 1996. Systematics of myliobatid elasmobranchs with emphasis on the phylogeny and historical geography of neotropical freshwater stingrays. (Potamotrygonidae: Rajiformes). Zoological Journal of the Linnaean Society 117(3):207–259.
Luer, C. A., and P. W. Gilbert. 1991. Elasmobranch fish-oviparous, viviparous, and ovoviparous. Oceanus 34(3):47–53.
Luer, C. A., P. C. Blum, and P. W. Gilbert. 1990. Rates of tooth replacement in the nurse shark, *Ginglymostoma cirratum*. Copeia 1990:182–190.
Maisey, J. G. 1979. Finspine morphogenesis in squalid and heterodontid sharks. Zoological Journal of the Linnaean Society 66:161–183.
———. 1984. Chondrichthyan phylogeny: a look at the evidence. Journal of Vertebrate Paleontology 4(3):359–371.
———. 1986. Heads and tails: a chordate phylogeny. Cladistics 21(3):201–256.
McEachran, J. D. 1970. Egg capsules and reproductive biology of the skate *Raja garmani* (Pisces: Rajidae). Copeia 1970:197–199.
———. 1977a. Reply to "Sexual dimorphism in skates (Rajidae)." Ecology 81(1):218–220.
———. 1977b. Variation in *Raja garmani* and the status of *Raja lentiginosa* (Pisces: Rajidae). Bulletin of Marine Sciences 27:423–438.

———. 1982. Chondrichthyes. Pp. 831–858 in Synopsis and classification of living organisms. McGraw-Hill, New York.

McEachran, J. D., and L. J. V. Compagno. 1982. Interrelationships of and within *Breviraja* based on anatomical structures (Pisces: Rajoidei). Bulletin of Marine Sciences 32(2):359–425.

McEachran, J. D., and K. A. Dunn. 1998. Phylogenetic analysis of skates, a morphologically conservative clade of elasmobranchs (Chondrichthyes: Rajidae). Copeia 1998(2):271–290.

McEachran, J. D., and J. D. Fechhelm. 1998. Fishes of the Gulf of Mexico. Vol. 1, Myxiniformes to Gasterosteiformes. University of Texas Press, Austin. 1,112 pp.

McEachran, J. D., and C. D. Martin. 1978. Interrelationships and subgeneric classification of *Raja erinacea* and *Raja ocellata* based on claspers, neurocranium, and pelvic girdles (Pisces: Rajidae). Copeia 1978(4):593–601.

McEachran, J. D., and R. E. Matheson Jr. 1985. Polychromatism and polymorphism in *Breviraja spinosa* (Elasmobranchii, Rajiformes). Copeia 1985:1035–1052.

McEachran, J. D., and T. Miyake. 1984. Comments on the skates of the tropical eastern Pacific: one new species and three new records (Elasmobranchii: Rajiformes). Proceedings of the Biological Society of Washington 17(4):773–787.

———. 1986. Interrelationships within a putative monophyletic group of skates (Chondrichthyes, Rajioidae, Rajini). Pp. 218–219 in Indo-Pacific fish biology: proceedings of the Second International Conference on Indo-Pacific Fishes. T. Uyeno, R. Arai, T. Taniuchi, and K. Matsuura (eds.). Ichthyological Society of Japan, Tokyo.

———. 1990a. Phylogenetic interrelationships of skates: a working hypothesis (Chondrichthyes, Rajoidii). Pp. 285–304 in Elasmobranchs as living resources: advances in the biology, ecology, systematics, and the status of the fisheries. H. L. Pratt Jr., S. H. Gruber, and T. Taniuchi (eds.). NOAA Technical Report NMFS 90.

———. 1990b. Zoogeography and bathymetry of skates (Chondrichthyes, Rajioidei). Pp. 305–326 in Elasmobranchs as living resources: advances in the biology, ecology, systematics, and the status of the fisheries. H. L. Pratt Jr., S. H. Gruber, and T. Taniuchi (eds.). NOAA Technical Report NMFS 90.

McEachran, J. D., and J. A. Musick. 1973. Characters for distinguishing between immature species of the sibling species *Raja erinacea* and *Raja ocellata* (Pisces: Rajidae). Copeia 1973(2):239–250.

———. 1975. Distribution and relative observation of seven species of skates (Pisces: Rajidae) which occur between Nova Scotia and Cape Hatteras. Fishery Bulletin 73(1):110–136.

McEachran, J. D., and M. Stehmann. 1977. Subgeneric placement of *Raja bathyphila* based on anatomical characters of the clasper, crania, and pelvic girdle. Copeia 1977(6):22–25.

———. 1984. A new species of skate, *Neoraja carolinensis*, from off the southeastern United States (Elasmobranchii: Rajoidei). Proceedings of the Biological Society of Washington 97(4):724–735.

McEachran, J. D., K. A. Dunn, and T. Miyaki. 1996. Interrelationships of the batoid fishes (Chondrichthyes: Batoidei). Pp. 63–84 in Interrelationships of fishes. M. L. J. Stiassny, L. R. Parenti, and G. D. Johnson (eds.). Academic Press, San Diego.

McFarlane, G. A., and R. J. Beamish. 1987. Validation of the dorsal spine method of age determination for spiny dogfish. Pp. 287–300 in Age and growth of fish. R. C. Summerfelt and G. E. Hall (eds.). Iowa State University Press, Ames.

McKenzie, M. D. 1970. First record of albinism in the hammerhead shark, *Sphyrna lewini* (Pisces: Sphyrnidae). Journal of the Elisha Mitchell Scientific Society 86(1):35–37.

McLellan, W. A., V. G. Thayer, and P. A. Pabst. 1996. Stingray spine mortality in the bottlenose dolphin, *Tursiops truncatus*, from North Carolina waters. Journal of the Elisha Mitchell Scientific Society 112(2):98–101.

Menni, R. C., and M. F. Stehmann. 2000. Distribution, environment, and biology of batoid fishes of Argentina, Uruguay, and Brazil: a review. Revue Museo Ciencias National, new ser., 2(2):69–109.

Merritt, N. R. 1973. A new shark of the genus (Squalidae, Squaloidea) from the equatorial western Indian Ocean, with notes on *Squalus blainvillei*. Journal of the Zoological Society, London 171:93–110.

Miller, W. A. 1995. Rostral dental development in sawfish (*Pristis perotteti*). Journal of the Aquariculture and Aquatic Sciences 7:98–107.

Mollet, H. F., G. M. Callet, A. P. Klimley, D. A. Ebert, A. D. Testi, and L. J. V. Compagno. 1996. A review of length validation methods for large white sharks, *Carcharodon carcharias*. Chap. 10 in Great white shark: biology of *Carcharodon carcharias*. A. P. Klimley and S. G. Ainley (eds.). Academic Press, San Diego. 33 pp.

Moore, C. J., and C. H. Farmer III. 1981. An angler's guide to South Carolina elasmobranchs. South Carolina Wildlife Marine Research Department, Charleston, S.C. 76 pp.

Moresi, L. R. 1957. The shark fishery industry. Ocean Leather Corporation, Newark, N.J. 12 pp.

Moss, S. A. 1967. Tooth replacement in the lemon shark, *Negaprion brevirostris*. Pp. 319–329 in Sharks, skates, and rays. P. W. Gilbert, R. F. Mathewson, and D. P. Rall (eds.). Johns Hopkins University Press, Baltimore, Md.

Mrosovsky, N. 2002. Hype. Marine Turtle Newsletter, no. 96:1–6.

Munoz-Chapuli, R., and F. Ramos. 1989. Morphological comparisons of *Squalus blainvillei* and *S. megalops* in the eastern Atlantic, with notes on the genus. Japanese Journal of Ichthyology 36(1):6–21.

Musick, J. A., and J. D. McEachran. 1969. The squalid shark, *Echinorhinus brucus*, off Virginia. Copeia 1969:205–206.

Myakov, N. A., and V. V. Kondyurin. 1986. Dogfishes, *Squalus* (Squalidae), of the Atlantic Ocean and comparative notes on the species of the genus from other regions. Journal of Ichthyology 26(6):1–8 (English); 26(4):560–574 (Russian).

Myrberg, A. A., and S. H. Gruber. 1974. The behavior of the bonnethead shark, *Sphyrna tiburo*. Copeia 1974:358–374.

Nakaya, K. K. 1973. An albino zebra shark, *Stegostoma fasciatum*, from the Indian Ocean, with comments on albinism in elasmobranchs. Japanese Journal of Ichthyology 20(2):120–122.

———. 1991. A review of the long-snouted species of *Apristurus* (Chondrichthyes, Scyliorhinidae). Copeia 1991(4):992–1002.

Nakaya, K. K., and A. Sato. 1997. Species grouping within the genus *Apristurus* (Elasmobranchii: Scyliorhinidae). Pp. 307–320 in Proceedings of the Fifth Indo-Pacific Fish Conference, Nouméa, New Caledonia, 3–8 November 1997. B. Seret and J.-Y. Sire (eds.). Société Franáaise d'Ichthyologie, Paris.

Nakaya, K. K., and M. Stehmann. 1998. A new species of deepwater catshark, *Apristurus aphyodes* n. sp., from the eastern North Atlantic (Chondrichthyes: Carcharhiniformes Scyliorhinidae). Archives of Fisheries Marine Research 46(1):77–90.

National Marine Fisheries Service. 1993. Fishery management plan for sharks of the Atlantic Ocean. United States Department of Commerce, Washington, D.C. 261 pp.

Naylor, G. J. H., and C. F. Marcus. 1994. Identifying isolated shark teeth of the genus *Carcharhinus* to species: relevance for tracking phyletic change through the fossil record. American Museum Novitates 3109:1–53.

Naylor, G. P. 1992. The phylogenetic relationships among requiem hammerhead sharks with inferring phylogeny when thousands of equally parsimonious trees result. Cladistics 8:295–318.

Naylor, G. P., A. P. Martin, E. B. Mattheson, and W. M. Brown. 1997. Interrelationships of Lamniform sharks: testing phylogenetic hypotheses with sequential data. Pp. 199–218 in Molecular systematics of fish. J. D. Kocker and C. C. Stephen (eds.). Academic Press, New York.

Nelson, J. S. 1994. Fishes of the world. 3rd ed. Wiley, New York. 600 pp.

Nevell, C. 1998. Bluntnose six-gilled shark. *Glaucus* 9(1):28–29.

Nishida, K. 1990. Phylogeny of the suborder Myliobatidodei. Memoirs of the Faculty of Fisheries, Hokkaido University, Hakodate, Japan 37(1–2):1–108.

Notobartolo-di-Sciara, G. 1987. A revisionary study of the genus *Mobula*

(Chondrichthyes, Mobulidae), with the designation of a new species. Zoological Journal of the Linnaean Society of London 91:1–91.

Notobartolo-di-Sciara, G., and E. V. Hillyer. 1989. Mobulid rays off eastern Venezuela (Chondrichthyes, Mobulidae). Copeia 1989:607–614.

Parker, H. W., and F. C. Stott. 1965. Age, size, and vertebral calcification in the basking shark, *Cetorhinus maximus* (Gunnerus). Zoologische Mededenlingen 40(34):305–319.

Parsons, G. R. 1985. Notes on the life history of the catshark, *Syliorhinus meadi*. Fishery Bulletin 83(4):694–695.

Pearce, A. S., and L. G. Williams. 1951. The biota of the reefs off the Carolinas. Journal of the Elisha Mitchell Scientific Society 67(1):133–161.

Pfeil, F. H. 1982. Tooth morphological examination of living and fossil sharks of the orders Clamydoselachiformes and Echinorhiniformes. Paleontology Zoology 1:315–318. T. H. Pheil, Munich, Germany.

Pratt, H. L., Jr., J. G. Casey, and L. B. Conklin. 1982. Observations in large white sharks, *Carcharias carcharias*, off Long Island, New York. Fishery Bulletin 80:153–157.

Pratt, H. L., Jr., C. McCandless, N. Kohler, C. Jensen, G. Ulrich, and N. Gulgan. 1998. Report of the 1998 Apex Predators Program, cooperative shark pupping and nursery grounds project. Apex Predators Program, United States DOC, NOAA, NMFS, NEFSC, Narragansett, R.I. 71 pp.

Purdy, R. W. 1996. Paleoecology of fossil white sharks. Pp. 67–78 in Great white sharks: the biology of *Carcharodon carcharias*. A. P. Klimley and D. G. Ainley (eds.). Academic Press, San Diego.

Purdy, R. W., V. P. Schneider, S. P. Applegate, J. H. McLellan, R. L. Meyer, and B. H. Slaughter. 2001. The neogene sharks, rays, and bony fishes from Lee Creek Mine, Aurora, North Carolina. Pp. 71–202 in Geology and paleontology of the Lee Creek Mine, North Carolina, III. C. E. Ray and D. J. Bohaska (eds.). Smithsonian Contributions to Paleobiology, no. 90.

Radcliffe, L. 1913. A summary of the work of the U.S. fisheries marine biological station at Beaufort, North Carolina during 1912. Science 38(977):395–400.

———. 1916. The sharks and rays of Beaufort, North Carolina. Bulletin of the United States Bureau of Fisheries 34:239–384.

Randall, J. 1973. Size of the great white shark, *Carcharodon*. Science 181:169–170.

Raschi, W., J. A. Munro, and J. L. V. Compagno. 1982. *Hypoprion bigelowi*, a synonym of *Carcharhinus signatus* (Pisces: Carcharhinidae), with a description of ontogenetic heterodonty in this species, and notes on its natural history. Copeia 1982(1):102–109.

Rider, S. J., M. Athorn, and G. O. Bailey. 2002. First record of a white tiger shark, *Galeocerdo cuvier*, from the northeastern Gulf of Mexico. Florida Scientist 65(1):13–15.

Rose, D. A. 1996. An overview of world trade in sharks and other cartilaginous fishes. TRAFFIC International, London. 106 pp.

Rosenberger, L. J. 2001. Phylogenetic relationships within the stingray genus *Dasyatis* (Chondrichthyes: Dasyatidae). Copeia 2001(3):615–617.

Rudloe, A. 1989. Habitat preferences, movement, size frequency patterns, and reproductive season of the lesser electric ray, *Narcine brasiliensis*. Northeast Gulf Science 10(2):103–112.

Sadowsky, V., and A. F. de Amorim. 1981. Occurrence of *Squalus cubensis* Rivero, 1936, in the western south Atlantic and incidence of its parasitic isopod *Lironica splendida* sp. n. Studies on Neotropical Fauna and Environment 126(3):137–150.

Sadowsky, V., C. A. Arfelli, and A. F. de Amorim. 1986. First record of the broadbanded catshark, *Etmopterus gracilipinis* Krefft, 1968 (Squalidae), in the Brazilian waters. Boletino Instituto Pesca 13(2):1–4.

Schwartz, F. J. 1959. White cownose ray, *Rhinoptera bonasus* (Mitchill), from Tangier Sound, Maryland. Maryland Tidewater News 15(3):12.

———. 1973. Spinal and cranial deformities in the elasmobranchs *Carcharhinus leucas*, *Squalus acanthias*, and *Carcharhinus milberti*. Journal of the Elisha Mitchell Scientific Society 89(1–2):74–77.

———. 1978. Body-organ weights for three basking sharks, *Cetorhinus maximus*, from North Carolina waters. Association of Southeastern Biologists Bulletin 25(2):29–48.

———. 1984. Sharks of the Carolinas. Institute of Marine Sciences, Morehead City, N.C. 52 pp.

———. 1989a. Occurrence, abundance, and biology of the blacknose shark, *Carcharhinus acronotus*, in North Carolina. Northeast Gulf Science 71:29–48.

———. 1989b. Sharks, sawfish, skates, and rays of the Carolinas. Special Publication of the Institute of Marine Sciences, Morehead City, N.C. 101 pp.

———. 1990. Mass migration congregations and movements of several species of cownose rays, genus *Rhinoptera bonasus*: a worldwide review. Journal of the Elisha Mitchell Scientific Society 106(1):10–13.

———. 1993a. Comments on selected parasites of elasmobranchs captured in the western Atlantic Ocean, especially off North Carolina, USA. Studia I Materialy Oceanografi 64. Marine Pollution 3:305–315.

———. 1993b. A North Carolina capture of the bramble shark, *Echinorhinus brucus*, family Carcharhinidae, the fourth in the western Atlantic. Journal of the Elisha Mitchell Scientific Society 104(3):158–162.

———. 1995. Elasmobranchs frequenting fresh or low salinities of North Carolina during 1971–1991. Journal of Aquariculture and Aquatic Sciences 7:45–51.

———. 1996a. Biology of the clearnose skate, *Raja eglanteria*, from North Carolina. Florida Scientist 59(2):82–95.

———. 1996b. Body-organ weight relationships of near-term and newborn tiger sharks, *Galeocerdo cuvier*, captured off North Carolina. Journal of the Elisha Mitchell Scientific Society 110(3–4):104–107.

———. 1998. History of the Poor Boy shark tournament in North Carolina in 1992–1996. Journal of the Elisha Mitchell Scientific Society 114(3):149–158.

———. 1999. Positive and negative outlook for United States and world shark fisheries. P. 836 in Proceedings of the Fifth Indo-Pacific Fish Conference, Nouméa, New Caledonia, 3–8 November 1997. B. Seret and J.-Y. Sire (eds.). Société Franáaise d'Ichthyologie, Paris.

———. 2000a. Elasmobranchs of the Cape Fear River, North Carolina. Journal of the Elisha Mitchell Scientific Society 116(3):206–224.

———. 2000b. Sharks new to the North Carolina elasmobranch fauna: the good, the bad, and the ugly. Journal of the Elisha Mitchell Scientific Society 116(3):292.

———. 2002. Basking sharks, *Cetorhinus maximus*, family Cetorhinidae, recorded in North Carolina waters, 1901–2002. Journal of the North Carolina Academy of Science 118(3): 201–206.

Schwartz, F. J., and G. H. Burgess. 1975. Sharks of North Carolina and adjacent waters. Information Services, North Carolina Department of Natural Economic Resources, Division of Marine Fisheries, Morehead City, N.C. 57 pp.

Schwartz, F. J., and M. D. Dahlberg. 1978. Biology and ecology of the Atlantic stingray, *Dasyatis sabina* (Pisces: Dasyatidae), in North Carolina and Georgia. Northeast Gulf Science 2:1–23.

Schwartz, F. J., and C. Jensen. 1995. Extreme habitat occurrences for two species of hammerhead sharks (family Sphyrnidae) in North Carolina and western Atlantic Ocean waters. Journal of the Elisha Mitchell Scientific Society 110(4):46–48.

Schwartz, F. J., and M. Maddock. 1986. Comparisons of karyotypes and cellular DNA content within and between major lines of elasmobranchs. Pp. 148–157 in Indo-Pacific fish biology: proceedings of the Second International Conference on Indo-Pacific Fishes. T. Uyeno, R. Arai, T. Taniuchi, and K. Matsuura (eds.). Ichthyological Society of Japan, Tokyo.

———. 2002. Cytogenetics of the elasmobranchs: genome evolution and phylogenetic implications. Marine Freshwater Research 53(2):491–502.

Schwartz, F. J., and H. Porter. 1977. Fishes, macroinvertebrates, and their ecological interrelationships with a Carolina scallop bed off North Carolina. Fishery Bulletin 75(2):424–446.

Schwartz, F. J., and G. W. Safrit Jr. 1977. A white southern stingray, *Dasyatis americana* (Pisces: Dasyatidae), from Pamlico Sound, North Carolina. Chesapeake Science 18(1):82–84.

Schwartz, F. J., W. T. Hogarth, and M. P. Weinstein. 1982. Marine and freshwater fishes of the Cape Fear estuary, and their distributions and relation to environmental factors. Brimleyana 7:17–37.

Schwartz, F. J., C. Jensen, and G. Hopkins. 1995. Occurrence of an adult male reef shark, *Carcharhinus perezi* (Carcharhinidae), off North Carolina. Journal of the Elisha Mitchell Scientific Society 111(2):121–125.

Scott, W. B., and S. N. Tibbo. 1968. An occurrence of the pelagic stingray, *Dasyatis violacea*, in the northwestern Atlantic. Journal of the Fisheries Research Board of Canada 25:1075–1076.

Sheehan, T. T. 1998. First record of the ragged-tooth shark, *Odontaspis ferox*, off the United States coast. Marine Fisheries Review 60(1):33–34.

Shirai, M. 1992a. Phylogenetic relationships of the angel sharks, with comments on elasmobranch phylogeny (Chondrichthyes, Squatinidae). Copeia 1992(2):505–510.

———. 1992b. A squalean phylogeny, a new framework of "Squaloid" sharks and selected taxa. Hokkaido University Press, Sapporo, Japan. 157 pp.

———. 1996. Phylogenetic interrelationships of neoselacheans (Chondrichthyes, Euselachii). Pp. 9–34 in Interrelationships of fishes. M. L. J. Stiassny, L. R. Parenti, and G. D. Johnson (eds.). Academic Press, San Diego.

Sims, D. W., and P. C. Reid. 2002. Congruent trends in long-term zooplankton decline in the northeast Atlantic and basking shark (*Cetorhinus maximus*) fishery catches off west Ireland. Fisheries Oceanography 11(1):59–63.

Slaughter, B. H., and S. Springer. 1968. Replacement of rostral teeth in sawfish and sawsharks. Copeia 1968(3):499–506.

Smale, M. J., and P. C. Heemstra. 1997. First record of albinism in the great white shark, *Carcharodon carcharias* (Linnaeus, 1758). South African Journal of Science 93(5):243–245.

Sminkey, T. R., and J. A. Musick. 1996. Demographic analysis of the sandbar shark, *Carcharhinus plumbeus*, in the western North Atlantic. Fishery Bulletin 94(2):341–347.

Smith, H. M. 1907. The fishes of North Carolina. Vol. 2. North Carolina Geological Economic Survey, Raleigh. 453 pp.

Smith, J. W., and J. V. Merriner. 1987. Age, growth, movements, and distribution of the cownose ray, *Rhinoptera bonasus*, in Chesapeake Bay. Estuaries 10(2):153–164.

Springer, S. 1939. Two new Atlantic species of dog sharks, with a key to the species of *Mustelus*. Proceedings of the United States National Museum 86(3058):461–468.

———. 1948. Oviphagous embryos of the sand shark, *Carcharias taurus*. Copeia 1948(3):153–157.

———. 1950a. Natural history notes on the lemon shark, *Negaprion brevirostris*. Texas Journal of Science 1950:349–359.

———. 1950b. A review of the North American sharks allied to the genus *Carcharhinus*. American Museum Novitates 1451:1–13.
———. 1959. A new shark of the family Squalidae from the Carolina continental slope. Copeia 1959(1):30–33.
———. 1960. Natural history of the Sandbar shark, *Eulamia milberti*. Fishery Bulletin 61:1–38.
———. 1963. Field observations on large sharks of the Florida-Caribbean region. Pp. 93–113 in Sharks and survival. P. W. Gilbert, J. A. F. Garrick, and L. P. Schultz (eds.). D. C. Heath, Boston.
———. 1966. A review of the western Atlantic catsharks, Scyliorhinidae, with descriptions of a new genus and five new species. Fishery Bulletin 65(3):581–624.
———. 1971. Three species of skates (Rajidae) from the continental waters of Antarctica. Biology of the Antarctic Seas 4. G. A. Llano and I. E. Wallen (eds.). American Geophysical Union. 10 pp.
———. 1979. A revision of the catsharks, family Scyliorhinidae. NOAA Technical Report NMFS Circular 422. 182 pp.
Springer, S., and H. R. Bullis Jr. 1956. Collections made by the *Oregon* in the Gulf of Mexico. U.S. Fish and Wildlife Service, Special Scientific Report on Fisheries 196. 129 pp.
Springer, S., and P. W. Gilbert. 1976. The basking shark, *Cetorhinus maximus*, from Florida and California, with comments on its biology and systematics. Copeia 1976:41–54.
Springer, S., and V. Sadowsky. 1970. Subspecies of the western catshark, *Scyliorhinus retifer*. Proceedings of the Biological Society of Washington 83(3):83–98.
Springer, V. G. 1964. A revision of the Carcharhinid shark genera *Scoliodon*, *Loxodon*, and *Rhizoprionodon*. Proceedings of the United States National Museum 115(3493):559–632.
Springer, V. G., and J. A. F. Garrick. 1964. A survey of vertebral numbers in sharks. Proceedings of the United States National Museum 116(3496):73–96.
Springer, V. G., and J. Gold. 1989. Sharks in question. Smithsonian Institution Press, Washington, D.C. 187 pp.
Stehmann, M. 1986. Notes on the systematics of the Rajid genus *Bathyraja* and its distribution in the world oceans. Pp. 241–268 in Indo-Pacific fish biology: proceedings of the Second International Conference on Indo-Pacific Fishes. T. Uyeno, R. Arai, T. Taniuchi, and K. Matsuura (eds.). Ichthyological Society of Japan, Tokyo.
———. 1999. Hare und Rochen in Gefahr Fischerei und Umvelteinfluse als Gefährdungsotentiale für Arten und Bestande, Bemühungen um Erhaltungsmasznahmen. Meer und Museum 15:67–72.

Stevens, J. D. 2000. The population status of highly migratory oceanic sharks in the Pacific Ocean: proceedings of the Symposium on Managing Highly Migratory Fishes of the Pacific Ocean, 4–6 November 1996, Monterey, California. National Coalition for Marine Conservation. Savannah.

Stevens, J. D., R. Bonfil, N. K. Dulvy, and P. A. Walker. 2000. The effects of fishing on sharks, rays, and chimaeras (Chondrichthyans), and the implications for marine ecosystems. ICES Journal of Marine Science 57:476–494.

Stiassny, M. L. J., L. R. Parenti, and G. D. Johnson. 1996. Phylogenetic interrelationships and neoselacheans (Chondrichthyes, Euselachii). Academic Press, San Diego. 496 pp.

Stillwell, C., and J. G. Casey. 1976. Observations on the bigeye thresher shark, *Alopias superciliosus*, in the western North Atlantic. Fishery Bulletin 74(1):221–225.

Taniuchi, T., and F. Yamagisawa. 1987. Albinism and lack of a second dorsal fin in an adult tawny nurse shark, *Nebrius concolor*, from Japan. Japanese Journal of Ichthyology 34(3):393–395.

Teaf, C. M., and T. C. Lewis. 1987. Seasonal occurrence of multiple caudal spines in the Atlantic stingray, *Dasyatis sabina* (Pisces, Dasyatidae). Copeia 1987(1):224–227.

Templeman, W. 1965. Some resemblances and differences between *Raja erinacea* and *Raja ocellata*, including a method of separating mature and large immature individuals of these two species. Journal of the Fisheries Research Board of Canada 22:899–912.

———. 1973. The skate *Raja richardsoni* Garrick, 1961, assigned to *Bathyraja*. Journal of the Fisheries Research Board of Canada 30(11):1729–1732.

Thies, D. 1987. Comments on Hexanchiform phylogeny (Pisces: Neoselachii). Zeitschrift für Zoologisches Systematic und Evolforschriften 25(3):188–204.

Thorson, T. B. 1973. Sexual dimorphism in numbers of rostral teeth of the sawfish, *Pristis perotteti* Müller and Henle, 1841. Transactions of the American Fisheries Society 102(3):612–613.

Van der Molen, S., G. Caille, and R. Gonzalez. 1998. By-catch of sharks in Patagonian coastal trawl fisheries. Marine Freshwater Research 49(7):641–644.

Walker, T. I. 1998. Can shark resources be harvested sustainably? A question with a review of shark fisheries. Marine Freshwater Research 49(7):553–572.

Weitzman, S. H. 1997. Systematics of deep-sea fishes. Pp. 43–77 in Deep-sea fishes. D. J. Randall and A. P. Farrell (eds.). Academic Press, San Diego.

Wilson, C. D., and M. P. Seki. 1994. Biology-population characteristics of *Squalus mitsukurii* from a seamount in the central Pacific area. Fishery Bulletin 92(4): 851–864.

Wilson, H. V. 1900. Marine biology at Beaufort. American Naturalist 43:339–360.

Wilson, P. C., and J. S. Beckett. 1970. Atlantic Ocean distributions of the pelagic stingray, *Dasyatis violacea*. Copeia 1970:696–707.

Yano, K. 1995. Reproduction biology of the black dogfish, *Centroscyllium fabricii*, collected from waters off western Greenland. Journal of the Marine Biological Association of the United Kingdom 78:285–310.

Yarrow, H. C. 1877. Notes on the natural history of Fort Macon, North Carolina, and vicinity, no. 3. Fishes. Proceedings of the Academy of Natural Sciences, Philadelphia 29:203–218.

Zangerl, R. 1973. Interrelationships of early chondrichthyians. Zoological Journal of the Linnaean Society of London 53(suppl. 1):1–14.

———. 1981. Chondrichthyes: 1. Paleozoic elasmobranchii. Pp. 1–115 in Handbook of Paleontology. Vol. 3a. Gustav Fischer Verlag, Germany.

index